Historical Geographies of Anarchism

In the last few years, anarchism has been rediscovered as a transnational, cosmopolitan and multifaceted movement. Its traditions, often hastily dismissed, are increasingly revealing insights which inspire present-day scholarship in geography. This book provides a historical geography of anarchism, analysing the places and spatiality of historical anarchist movements, key thinkers, and the present scientific challenges of the geographical anarchist traditions.

This volume offers rich and detailed insights into the lesser-known worlds of anarchist geographies with contributions from international leading experts. It also explores the historical geographies of anarchism by examining their expressions in a series of distinct geographical contexts and their development over time. Contributions examine the changes that the anarchist movement(s) sought to bring out in their space and time, and the way this spirit continues to animate the anarchist geographies of our own, perhaps often in unpredictable ways. There is also an examination of contemporary expressions of anarchist geographical thought in the fields of social movements, environmental struggles, post-statist geographies, indigenous thinking and situated cosmopolitanisms.

This is valuable reading for students and researchers interested in historical geography, political geography, social movements and anarchism.

Federico Ferretti is a Lecturer in Human Geography in the School of Geography, University College Dublin (UCD), Ireland.

Gerónimo Barrera de la Torre is a graduate student at the University of Texas at Austin, USA.

Anthony Ince is a Lecturer in Human Geography in the School of Geography and Planning, Cardiff University, UK.

Francisco Toro is a Lecturer in Human Geography at the Department of Regional Geographical Analysis and Physical Geography, University of Granada, Spain.

Routledge Research in Historical Geography

Series Edited by Simon Naylor (School of Geographical and Earth Sciences, University of Glasgow, UK) and Laura Cameron (Department of Geography, Queen's University, Canada)

This series offers a forum for original and innovative research, exploring a wide range of topics encompassed by the sub-discipline of historical geography and cognate fields in the humanities and social sciences. Titles within the series adopt a global geographical scope and historical studies of geographical issues that are grounded in detailed inquiries of primary source materials. The series also supports historiographical and theoretical overviews, and edited collections of essays on historical-geographical themes. This series is aimed at upper-level undergraduates, research students and academics.

For a full list of titles in this series, please visit www.routledge.com/Routledge-Research-in-Historical-Geography/book-series/RRHGS

Published

Historical Geographies of Prisons: Unlocking the Usable Carceral Past
Edited by Karen Morin and Dominique Moran

Historical Geographies of Anarchism: Early Critical Geographers and Present-Day Scientific Challenges
Edited by Federico Ferretti, Gerónimo Barrera de la Torre, Anthony Ince and Francisco Toro

Historical Geographies of Anarchism

Early Critical Geographers and Present-Day Scientific Challenges

Edited by
Federico Ferretti, Gerónimo Barrera de la Torre, Anthony Ince and Francisco Toro

LONDON AND NEW YORK

First published 2018 by Routledge

2 Park Square, Milton Park, Abingdon, Oxfordshire OX14 4RN

52 Vanderbilt Avenue, New York, NY 10017

Routledge is an imprint of the Taylor & Francis Group, an informa business

First issued in paperback 2019

British Library Cataloguing in Publication Data
A catalogue record for this book is available from the British Library

Library of Congress Cataloging in Publication Data
A catalog record for this book has been requested

ISBN: 978-1-138-23424-6 (hbk)
ISBN: 978-0-367-21900-0 (pbk)

Typeset in Times New Roman
by Taylor & Francis Books

Contents

Illustrations

Contributors

Narciso Barrera-Bassols is Professor at the Universidad Autónoma de Querétaro, México. Trained in Anthropology, Narciso is now a specialist of Environmental Geography and member of a number of Latin American groups of socially engaged scholars.

Gerónimo Barrera de la Torre is PhD Student at the University of Texas at Austin. After completing his research master in Mexico, Gerónimo is now working on post-statist geographies and postcolonial anthropologies.

Julian Brigstocke is Lecturer in Human Geography at the University of Cardiff. Julian is a cultural geographer and social theorist interested in the relationships between geography and aesthetics, with a special focus on the role played by power and authority.

David Crouch is Professor of Geography at the University of Derby. Interested in cultural geography, social anthropology, cultural and visual studies and art theory, David collaborated with the famous urbanist and anarchist thinker Colin Ward.

Federico Ferretti is Lecturer in Human Geography at University College Dublin. Trained as an historical geographer, Federico works on the international networks of early anarchist geographers and on anti and postcolonial geographies with a special focus on Latin America.

Andrew Hoyt, historian, is about to conclude his PhD dissertation at the University of Minnesota, focusing on transatlantic radical print culture, particularly the poetry and art embedded in Italian language anarchist publications (1880s to 1940s).

Anthony Ince is Lecturer in Human Geography at the University of Cardiff. His interests are situated in the intersections of political and social geographies and apply to a wide range of empirical subjects, including far-right political movements, backpacking, labour, and the so-called 'sharing economy'.

Carl Levy is Professor of History at Goldsmiths University, London. A specialist in the history of Italian anarchism and especially in the figure of Errico

Malatesta, Carl is interested in the wider field of comparative labour history in Europe.

Pascale Siegrist, historian, is about to conclude her PhD dissertation at the University of Konstanz. Trained in Russian Studies, Pascale works on the international networks of anarchist geographers Kropotkin and Reclus.

Simon Springer is Associate Professor of Geography at the University of Victoria. A specialist in political violence in Cambodia, he is one of the most famous international protagonists of present-day revival of anarchist geographies.

Francisco Toro is Lecturer in Human Geography at the University of Granada. Francisco's research interests lie in sustainability, ecological crisis, planning, and the nature–society interactions.

Davide Turcato was granted his PhD at the Simon Fraser University Vancouver. Historian, he is an internationally recognized specialist of the transnational networks of Italian anarchism and editor of the monumental project of the Complete Works of Errico Malatesta.

Rita Velloso is Professor of Architecture and Urbanism at the Federal University of Minas Gerais and at the Catholic University in Belo Horizonte. Her research interests lie in urbanism, social movements and critical ideas on the city, with a special focus on the relations between architecture and philosophy.

Foreword: Anarchy is forever: The infinite and eternal moment of struggle

Simon Springer

> Life exists only at this very moment, and in this moment it is infinite and eternal, for the present moment is infinitely small; before we can measure it, it has gone, and yet it exists forever.[1]

Geography has a very unpleasant history. From the very outset of the discipline, geography was at the frontlines of the colonial project. It continues to be entangled with militarism through a steady stream of funding for geographers who see no ethical qualms about the perpetuation of war. The Bowman Expeditions immediately come to mind, where geography professors, Jerome Dobson and Peter Herlihy, from the University of Kansas took funding from the U.S. Army's Foreign Military Studies Office to map indigenous lands in Mexico. Ostensibly this project sought to advance the rights of indigenous peoples, yet the researchers failed to disclose who was bankrolling the project to participating communities and the fact that the collected data was being transmitted to Radiance Technologies, a military contractor that billed itself as 'creating innovative solutions for the warfighter'. The project culminated in controversy, as the Zapotec people discovered they had been duped.[2] This was undeniably a very ugly moment for geography, bringing tremendous shame and embarrassment to the discipline. But what might happen if we start thinking about geography in a different light, where instead of paying service to military pursuits and ceding to the interests of imperialists, we were to instead orient it towards the anarchist horizons of possibility?[3]

This is the exact question that this book grapples with and it is a concern that Pëtr Kropotkin and Élisée Reclus were already raising over a century ago. While their contemporaries were busy contributing to the realisation of colonial machinations and planning for the next round of stripping indigenous peoples of their lands,[4] these two men were tearing apart the racist underpinnings of environmental determinism and eviscerating all the implications that were being assumed of Charles Darwin's work. Rather than allowing geography to be a means of justifying domination and war, Reclus and Kropotkin viewed the discipline as a conduit for dissipating prejudice by aligning it to anarchism. This desire for an anarchist geography should not be thought of as a quaint and idiosyncratic memory of the past, subsumed beneath the tides of more important geographical traditions. Instead, there is much to learn from anarchist geographies today, where anarchism pulses within many of the discipline's current veins. Reclus'[5] notion of a 'universal geography' for example has significant resonance with the recent relational turn in geography, and his environmental outlook anticipated the field of political ecology. Similarly, Kropotkin's[6] theory of 'mutual aid' is a primary source of inspiration for emergent ideas around reciprocity and the geographies of care, but it also hits at the possibilities of current more-than-human geographies by recognising the agency of non-human actors and the symbiotic relationships that enable our very existence on planet Earth. A reengagement of anarchism within contemporary geographical theory and practice opens a renewed path towards the open fields of emancipation, veering far to the left of the thorny thickets of militarism and imperialism that sadly some geographers continue to preen. Yet in realising the possibilities that can grow out of the synergies between anarchism and geography within the present moment, it is important to look to the lessons of the past.

We can start our journey along the contours of anarchist geographies in the depths of time immemorial. It is in the unknowable mystery of this olden domain that anarchy first took shape as the very condition of life itself. Anarchy is indicative a world free from servitude and the intrusions of governance, where there are no hierarchical institutions or mechanisms of control. It reveals a world of free association and constant change, a deep interdependence between everything that exists and the perpetual evolution of the unfolding interactions of being. In the midst of this immanence, domination represents a disruption, where in the grand scheme of things it is quickly corrected by the prevailing order of existence. Anarchy can accordingly be considered at once the infinite and unfathomable vibrations of the universe and the geometry of life itself. It is the energy that flows through our natural world, a beautiful fractal that can never actually be broken, only temporarily interrupted. Any suggestion that anarchy is chaos consequently signals a profound lack of understanding and serves not as truth, but as the manifestation of an anxiety born from the parochial hubris of the human mind as it attempts to restructure what actually exists in accordance with what it problematically thinks should be. Anarchy is only mayhem through the distorted lens of a fool's sense of order.

In contrast to anarchy, anarchism is a political philosophy and practice that attempts to correct the strange intermission of the present moment, where the rhythm and flow of symbiosis has been disrupted by extraordinary mass violence. The state, capitalism, religion, sexism, racism, ableism, childism, and speciesism are all representative of the *archy*, or systems of domination, that form the nebula of this interference, clouding our vision through the myopia of gradation and supremacy. It is precisely these facets of rule that are the target of anarchists. In attempting to realise the end of such cruelty, anarchists recognise that there is no primacy to the ordering of life, only the harmony of oneness. We are connected to existence as equals, with none taking precedence over another. On a larger temporal timescale it is guaranteed that the system will correct itself, whereby consonance will be restored and all existent chains of command severed. Greed, extractivism, and the accumulation of capital will push humanity to extinction and the entire order of our planet will reset itself, shattering the false dichotomies and hierarchies that humans have fabricated. Yet for anarchists this is not the desired outcome. The pursuit of anarchism is an attempt to restore balance to the world before our collective demise becomes assured. It is a reaction to the nihilism of avarice, premised on the very radical idea that humans should continue to be woven into the fabric of the great unravelling enigma that is the universe.

Through the institution of the state and the spread of capitalism we have collectively wrestled with the natural world, imposing hierarchies and modes of domination upon the structure of our planetary existence that simply don't make sense. They are the reflections of human arrogance that have taken us through the looking glass into a dystopian reality of profound malevolence. We can take some measure of solace in the idea that the state and capital are facile and fugitive attempts at organisation that will undoubtedly come undone, but any politics of resignation is fundamentally a practice of necromancy. These are institutions that signify the celebration of our demise and any communion with them is the fulfilment of a death wish. If we do nothing and simply wait for the eternal recurrence to arrive,[7] our shared misery as we plunge headlong towards oblivion is virtually assured. Anarchism requires more of us. An anarchist politics insists that apathy give way to empathy. It demands the impossible, summoning us to take action in recreating the world through the realignment of our geographies towards the possibility of a tomorrow that at present cannot be assured. The primary mode of restitution is prefiguration, a process of living life today, *in this very moment, in this exact space*, in a way that is befitting of the future we seek to establish.[8] Our future is consequently to be found in the past, in the primordial anarchy that is ancient beyond memory, record, or tradition, and the living anarchism that was documented in historical struggles.

The chapters that comprise this book offer a powerful reminder of what we can jointly achieve when we are willing to struggle in the face of oppression. The anarchist lessons of the past are brought to bear on geography, not as the anachronistic stains of a yesterday that can never be revisited, but as a vital pigmentation of what becomes possible today when we have the courage to see the full spectrum of colour that this life provides. Just as anarchy was the stuff of *there* and *then*, anarchism attempts to realise a *here* and *now*. Yet this too is indicative of a false separation of space and time, demonstrating the ways that language all too often fails us. *There* is *here*, and *then* is *now*. Anything else is illusion, veiled in the ignorance of separation. The cosmology of space-time folds into itself, and the *eternal-now-infinite-here* is but a matter of fact, the very basis of existence.

So if anarchy is the filament of our past, then anarchism is the incandescence of our present. Each plays a role in illuminating our future with the passionate radiance of connection. If we are to remain a part of the chronicle of life in its beautiful mercurial narration, we need to let the *stories-so-far* of anarchism guide us into a future that embraces our past as the space of anarchy. It is the embrace of this *infinite and eternal moment of struggle* that sees us emerge from our chrysalis to spread our wings. It gives purpose to the work of transforming the world as we glide along a cyclical line of flight towards the reawakening of harmony. Anarchism is an uroboric geography. It is the realisation of the macrocosm in the microcosm, the momentary in the infinite, the universe in a speck of dust. Through explorations of the past, the musings offered in the pages of this book promote a vision of the horizon, enabling us to realise that anarchist geographies are the fulfilment of a world that we have the active ability to create, an ontology that yearns to be created. They envision an earth beyond militarism, beyond cruelty, beyond violence, and beyond hate, in short, a world that we would actually want to live in. They remind us that while anarchism may be fleeting, anarchy is forever.

Notes

1 Watts, *Become What You Are*, 10.
2 Bryan and Wood, *Weaponizing Maps*: Indigenous Peoples and Counterinsurgency in the Americas.
3 Springer, *The Anarchist Roots* of Geography: Toward Spatial Emancipation.
4 Driver, *Geography Militant*: Cultures of Exploration and Empire.
5 Reclus, *The Earth and Its Inhabitants*: The Universal Geography.
6 Kropotkin, *Mutual Aid*: A Factor of Evolution.
7 Nietzsche, *Thus Spake Zarathustra*.
8 Ince, 'In the Shell of the Old: Anarchist Geographies of Territorialisation'.

Introduction

Federico Ferretti, Gerónimo Barrera de la Torre,
Anthony Ince and Francisco Toro

In the last few years, anarchism has been rediscovered as a transnational, cosmopolitan and multifaceted movement and its traditions, often hastily dismissed in the name of Marxism, Liberalism or post-structuralism, are increasingly revealing insights which inspire present-day scholarship in anthropology, sociology, philosophy, biology, social history and, last but not least, geography. This work is the first attempt to provide a historical geography of anarchism, addressing at the same time places and spatiality of historical anarchist movements and key thinkers, and the present scientific challenges of the geographical anarchist traditions in the fields of social movements, environmental struggles, post-statist geographies, indigenous thinking and situated cosmopolitanisms.

This book collects the proceedings of the multiple session 'Historical Geographies of Anarchism: Situating Struggles, Studying Environments' organised by the Editors for the Royal Geographical Society–Institute of British Geographers Annual International Conference on the theme of 'Geographies of Anthropocene', which took place in Exeter in September 2015. While in recent years there has been a growing number of sessions on anarchism's relationship with geography, this conference was the first one organised on these specific historical topics and the contributions drew on three main strands of international literature on Anarchist geographic research.

The first strand concerns the recent rediscovery of anarchist geographies, which has occurred increasingly since the special issues consecrated to anarchist geographies by the journals *ACME* and *Antipode* in 2012. These journal issues included works on anarchist geographies' genealogies[1] and historical figures like Léon Metchnikoff/Lev Mečnikov (1838–1888), Élisée Reclus (1830–1905)[2] and Pëtr Kropotkin (1842–1921)[3] now addressed by a great number of multilingual contributions.[4] Recent works have also promoted new debates in the fields of anarchist pedagogies[5] and post-statist geographies,[6] among numerous other critical, substantive contributions to existing debates. The second is the literature analysing historical geographies of science and scientific revolutions[7] as a part of the wider context of the spatial turn in social sciences, addressing the localisations and mobilities of scientific knowledge as decisive elements in understanding it.[8] The third strand of the book is the historical literature considering anarchism as a transnational movement based on networks and cosmopolite circulations of ideas, publications and militants,[9] a line of research which has successfully challenged the shortcomings of more traditional histories of anarchism which were dependent on what Davide Turcato defines as a 'cyclical pattern of advances and retreats', leading to false 'millenaristic' readings of anarchism that impede a clear understanding of how this movement really worked.[10]

A common topic for a great part of our contributions is an analysis of the transnational and cosmopolitan networks and circulations affecting both social movements and the construction of knowledge by early anarchist geographers: the transnational nature of the anarchist movement can help to explain the anti-colonialist thinking of its early intellectuals, such as Reclus and Kropotkin. Thus, the idea of linking anarchism and its history to its places and circulations is a central one for this collective work, which addresses at the same time the history of anarchism and present-day anarchist movements and their spatiality. Consistent with the transnationalism, cosmopolitism and multilingualism of the anarchist tradition, we include works produced by an international collective of authors who worked (at the date of the conference) in Brazil, Mexico, Spain, Germany, Sweden, Ireland, Switzerland, UK and USA. Thus, one of our aims is to give accounts of works and debates outside the English-speaking world, which has become the *de facto* centre of academic knowledge production, and beyond mainstream academia altogether. Another characteristic of this book is its interdisciplinary nature, as one can find among the authors, historians, historical geographers, cultural and political geographers who have found in the spatiality of anarchism a common ground for research and discussion.

The first part of the book addresses spaces and places in the history of anarchism under a transnational standpoint. Carl Levy traces a wide historical outlook of the city as the place for anarchist experiments in self-government and as a generator of powerful revolutionary imaginaries. The historical experiences raised in this essay include the medieval city, the 1871 Paris commune, the 1936 revolutionary Barcelona and the capitals of 1968 risings, interrogating each case on its significance for the history and development of anarchist thinking. Andrew Hoyt presents a geographical approach to the distribution of the Italian-speaking anarchist journal *Cronaca Sovversiva* all over the United States, in order to analyse the spatial patterns of distribution of transnational anarchist propaganda by applying the concept of 'social field'. Davide Turcato likewise addresses the Italian anarchist press in the USA, in this case to investigate the relation between internationalism, cosmopolitan practices and ideas of national cultural belonging among transnational anarchist militants. Julian Brigstocke analyses a body of documents from late nineteenth century France to open up a discursive critique of the relationship between humour and violence in militant mentality and in wider biopolitical practices of that time.

The second part of the book analyses topics related to the spaces of early anarchist geographers, focusing mainly on the figures of Reclus and

Kropotkin. These authors were the inventors of a solidaristic interpretation of evolution, known as the theory of mutual aid, which challenged established theories of the time, such as Malthusianism and Social Darwinism. At the same time, they were committed to the study of the ongoing relations between humankind and environment, refuting the positioning of so-called 'human' and 'natural' environments as separable domains. Drawing less on ideas of wilderness and protection, these thinkers sought solutions to 'harmonise' the coexistence of living beings on the earth's surface, anticipating some of the contemporary themes of more-than-human approaches. Secular and rational 'science' was thus considered as a fundamental instrument for that. Problematising all these topics entails also reflexions and new debates on the present coming back of Creationism, Malthusianism and environmental Determinism. Francisco Toro analyses the Reclus's thinking with regard to the intellectual context of present-day degrowth theories, showing Reclus's commitment lead to very effective concerns in today's fields of geography of resources and environmental geographies. Federico Ferretti addresses three cases of analysis by early anarchist geographers Reclus, Kropotkin and Mikhail Dragomanov [or Drahomanov] (1841–1895) of colonised or recently decolonised nations in order to understand the complex links between anarchism, nationalism and anti-colonialism in the Age of the Empire. Pascale Siegrist analyses Kropotkin's and Reclus's commitment to the scientific field of their time in order to problematise the definition of 'anarchist geography', a relatively recent label, which did not exist during their period of activity.

The third part of the book addresses the spatiality of present challenges for anarchist geographers. David Crouch analyses the effectiveness of the social geography of Colin Ward (1924–2010), a thinker who is considered as one of the most important references for present-day anarchism, mainly in English-speaking countries. Crouch addresses at the same time his personal experience of collaboration with Ward, the latter's references in the history of geographical thought and his insights for contemporary cultural geographies and urban planning. Rita Velloso analyses the spatiality and 'insurrectionary architecture' of the 2013 'Brazilian Spring' as it occurred in Belo Horizonte, its relations with urban spaces and the general social aims of the movement. Anthony Ince and Gerónimo Barrera address an innovative linkage between non-statist geographies and de-colonial geographies, matching in this sense the historical tradition of anarchist geographers Reclus and Kropotkin, committed to building a regional science which did not assume the state and the administrative boundaries as its framework of reference. For today's critical and radical geographies, the association with archaeology means a deeper historical reflexion in thinking and questioning recent assumptions and discourses on the dissolution of states, emphasising the necessity for critical scholarship to destabilise the state as an analytical category. Finally, Gerónimo Barrera de la Torre and Narciso Barrera-Bassols analyse the relations between anarchism and indigenous movements, mobilising an anarchist view of Anthropocene which draws on Reclus's idea of the consubstantiality of the terms traditionally considered as 'humankind' and 'nature'. The authors propose an analysis of contemporary ecological emergencies through the lens of a set of perspectives distinct from the classical European and Eurocentric intellectual tools. In particular, they utilise indigenous thinking and seek to integrate it with a range of anarchist ideas on plurality and social solidarity. Thereby, they target an alternative view of modernity, drawn from what de-colonial thinkers in Latin America call 'pluriversality'.[11]

These contributions present a heterogeneous panorama in their historical and geographical span, but they also share important points in common. First, their common understanding of anarchism as a transnational, cosmopolitan and multilingual tradition which has to be studied in its places, networks and flows. Second, is the importance of the historical tradition of the 'classical' anarchist geographers to provide insights for present day revival of anarchist scholarship in geography: this legacy, as several of our authors argue, should not be taken uncritically, but to be first historically and spatially contextualised to interact with recent debates while avoiding anachronism. Third, there is a common commitment to rethink the epistemological framework of geography questioning the long-lasting hegemony of readings based on the state, or on state-like territories, as the privileged framework of reference. Networks, material and immaterial flows, diverse localisations of intellectual and political movements addressed in this book show how much more complex the spatialities of these phenomena are; likewise, present-day global and reticular protest movements all over the world, and the left/libertarian revolutions which occurred in regions like Chiapas and more recently in Rojava,[12] show that state and state reason are increasingly intellectually unfit to explain reality and to inform scholarship committed to social transformation. Through the contributions of this book and its authors, we therefore seek to open up new analytical frameworks and frontiers of study for the future development of historical geographies in general, and anarchist historical geographies in particular.

Notes

1 Springer, *The Anarchist Roots of Geography*.
2 Ferretti, *Élisée Reclus*; Pelletier, *Géographie et anarchie*.
3 Adams, *Kropotkin*; Kinna, *Kropotkin*; MacLaughlin, *Kropotkin*.
4 For an idea of the international and multilingual contributions consecrated to these figures, consult the site http://raforum.info/reclus/
5 Springer, White and Souza, *The Radicalization of Pedagogy*.
6 Ince and Barrera, 'For Post-Statist Geographies'.
7 Livingstone and Withers, *Geography and Revolution*; Livingstone and Withers, *Geographies of Nineteenth-Century Science*.
8 Naylor, 'Historical Geography'; Secord, 'Knowledge in Transit'.
9 Bantman, *The French Anarchists*; Di Paola, *The Knights Errant*; Hirsch and van der Walt, *Anarchism and Syndicalism*.
10 Turcato, 'Italian Anarchism as a Transnational Movement', 408.
11 Carrillo, *Pluriverso. Un ensayo sobre el conocimiento indígena contemporáneo*.
12 Knapp, Flach and Ayboga, *Revolution in Rojava*.

PART I

Spaces of the history of anarchism

1 Anarchists and the city

Governance, revolution and the imagination

Carl Levy[1]

Since the emergence of classical anarchism in the mid-nineteenth century, the city and the urban commune have been central to the anarchist imagination and anarchist socio-political action. This chapter presents a synoptic overview of the uses of the city in the anarchists' programmes, tactics, strategies and visions. From the Paris Commune of 1871, as the symbol of the revolution, to the role of the anarchists in Barcelona during the Spanish Civil War. Although I do not discuss the role of the city in anarchist thought and practice after 1945, the message is clear: the city has been central for the transformation of philosophical anarchism into a quotidian, vivid practice from the 1860s to the present day.[2]

Introduction: anarchism and the city, the context of the argument

Before there was a movement of self-declared anarchists, the first 'anarchists' were called Mutualists, Federalists and Internationalists. Just as Marxism as an ideology evolved into a corpus of academic and doctrinal statements and programmes, and assumed a public face in the late nineteenth century, so too anarchism, in reaction to Marxism, but also in reaction to events on the ground, became a self-contained identifiable ideology and movement only in the 1880s and 1890s.[3] For the advocates of anarchism, decentralised power structures in towns and cities were used to galvanise the imagination and the movements for the final goals of a stateless and anti-authoritarian world.

The study of the relationship of the anarchists to the city is useful in two regards. It helps to bridge the gaps and the controversies over the periodisation of anarchism (as a formal ideology) and its precursors, namely pre-anarchism, classical anarchism (1860s to 1945), and new and post-'anarchisms' (1945 to the present): thus for example Pëtr Kropotkin invoked aspects of the late medieval European city-state as a model for his modern anarchist city of 1900 and Colin Ward invoked libertarian solutions for the London of the 1960s and 1970s by invoking Kropotkin.[4] This anarchist/city optic also is useful in the vexed discussion of whether or not anarchism was just another European provincial or Orientalist ideology, which accompanied the steamship, the telegraph, the missionary and the machine gun. To what extent did the arrival of anarchism and syndicalism in Latin America, China, Japan or India feed off indigenous forms of thought and action and to what extent, as has been shown recently in the case of Japan, where Russian Populist progenitors were inspired by non-Western models of cooperation in civil society, did non-European forms of libertarian anarchists inspire European anarchists and anarchism?[5] How can we imagine Classical Anarchism without taking into account the vibrant movements of Argentina, Cuba or Mexico?[6] One way to address these issues is through examining movements found in the liminal cities of the Global South during the era of High Imperialism (1880–1920), thus for example, Buenos Aires, Shanghai, Havana or Beirut as well as the liminal cities of the Imperial overlords from San Francisco to London to New York to Barcelona.[7] Not only does the study of the theme of anarchism and the city challenge the accepted chronology, it can also serve as a methodological tool, which grounds the recent interest in transnational, cosmopolitan and network approaches in a solid, day to day reality of the urban milieu, which can be grasped by the historian and also by social and political scientists who study the dissemination and mutation of political ideologies and political practices.[8]

Unlike other political movements, the study of anarchism relies upon ground-level social history to understand its nuances and continuities because long-term forms of organisation can be elusive or short-lived. Thus, to quote, Tom Goyens, in his suitably entitled monograph (*Beer and Revolution*) a study of the German anarchists in New York City from 1880 to 1914, that particular centre of conviviality, the beer hall.[9]

> A social and cultural history of German anarchists in the greater New York area must take into account the geography of the movement, its physical connection to the urban landscape. This movement was not merely an intellectual phenomenon, or some elusive threat – the ghost of anarchy – in the minds of respectable citizens. It consisted of men, women and children of exiled and immigrated families, of impetuous activists who were part of the citizenry of New York.

Anarchism became flesh and punched over its weight, through global syndicalism, in counter-institutions such as free schools and social centres, and in the tissues of diasporic and immigrant communities, such as the Italian colony of London (1870–1914), studied by Pietro Di Paola[10] or the contemporaneous French colony, brought to life by Constance Bantman.[11] Studying the role of the city, I think, is a red thread, which joins together syndicalism, conviviality and educational institutions. But let me add a disclaimer. I am not arguing that other approaches are not important: the studies of rural movements of the Zapatistas of the Mexican Revolution or the movement of the Maknovscina in rural Ukraine during the Russian Civil War (1918–1921),[12] or even the ground-breaking and delightful work of the anthropologist James Scott, who identified a zone of anarchist-like structures and behaviours in upland South-East Asia (Zomia)[13] in the early modern and the initial part of the modern eras, and the works on pirate confederacies and maroon settlements in the Americas, are all significant to our understandings of anarchism.[14] But in this chapter, I will show that the urban optic has its utility.

The Commune of Paris 1871 and its repercussions

The Commune of Paris of 1871 lasted just 72 days but it became the focus for the imaginations of Karl Marx, Michael Bakunin, Vladimir Lenin, William Morris, Pëtr Kropotkin, Louise Michel and Élisée Reclus. The city of Paris was abandoned by the provisional government after the defeat by the Prussians, and the radicals of Paris, stirred by the denizens of the popular and working-class clubs, which had flourished since the late 1860s during the liberalisation of politics in the waning days of rule of Louis Napoleon, took control. The politics and policies of the Commune were marked by improvisation but the central themes were clear: the Universal Republic, a France of decentralized political units and a Paris ruled in turn by its arrondissements (the Central Committee of the Twenty Arrondissements). Public policies announced the institutions of free secular education for all children: a polytechnic education, which combined manual and intellectual training, but also a system of crèches for younger children. The renter would win out over the landlord. Women were noticeably present in this polity, with the Women's Union the largest and most effective institution of the Commune. As Kristin Ross notes, artists were a predominant force in the Commune – the painter Courbet was joined by a legion of decorative artists and the practitioners of woodworking and shoe-making. Art was to be universal and not imprisoned in the Salon. The anarchist Reclus proclaimed that aesthetic concerns were also concerns of the democratic polity, and thus heralded the birth of a communal luxury based on the 'principles of association and cooperation'.[15] But the Commune was a balance between reformists and revolutionaries, it even contracted a loan from the Rothschilds and reassured lenders that debts would be repaid and it never seized the funds of the Bank of France. But it also outlawed night work in bakeries and created worker controlled munition shops to arm the National Guard.

The lessons from the Commune were varied. For Lenin, the Commune-State needed a vanguard party to protect it from counter-revolution, while, it is argued by Ross and others that in his last decade of his life, Marx used the example of the Commune of Paris to soften his hostilities to communal socialism of the agrarian Populists of Russia. For the anarchists in the late nineteenth century, the Commune of Paris was the turning point for the formation of their ideology and a counterpoint to authoritarian Marxism. Indeed Bakunin's collectivist anarchism was crystallised here: it was the transformation of the Universal Republic into a quest for the realisation of internationalist feder-alism. But although the Commune was celebrated and grieved by the anar-chists in the late nineteenth century, it was not beyond their criticism.[16] Thus Errico Malatesta thought its social policies had been too restricted and timid, that it was evolving disturbingly towards representative rather than direct democracy, and that the more dictatorial Jacobins and Blanquists had become too powerful in the Committee of Safety.[17] William Morris and Pëtr Kropotkin would join the imagery of the Commune with a critical reinterpretation of modernity and particularly the growth of great capitalist conurbations such as London.[18] In Morris's *News from Nowhere* the insurgents of the English Revolution echo the levelling of the Napoleonic Victory Column in the Place Vendôme by the Communards by turning Trafalgar Square into an orchard.[19]

But Morris had little time for the industrial city as such, whereas Kropotkin combined the praise for anachronisms – the so-called primitive forms of democracy and adaptive cooperation, the mutual aid carried out by peasants, farmers, and the First Nations of the harsh landscapes of Siberia and the freeholders of Iceland (this of course shared with Morris) with an appreciation of the modern city. But the most recent and more modern might not necessarily be more evolved and better than past models of governance. For Kropotkin there were two roads in the history of Europe: the road traversed which embraced the communal liberties of the city states and the urban guilds of late medieval Europe, that is in the era previous to the rise of the Absolutist State, and a Roman imperial road which led to his troubled present-day of militarist imperial states of Europe, with their centralizing monster capital cities.[20] But the other road from the city-state demonstrated Kropotkin's creative use of the anachronism, and his approach was different from Morris's or the Russian Populists', who were partial to the smaller settlement of the *mir*. For Kropotkin the decentralised city, based on the high technology of his day – electricity – would humanise modernity, by the interweaving of fields, factories and workshops, through Garden Cities, promoted by his followers such as Patrick Geddes.[21] Here is another follower, Lewis Mumford in 1961, but it could easily be 2011:

> Almost half a century in advance of contemporary opinion (Kropotkin) had grasped the fact that flexibility and adaptability of electronic communication and electric power, along with the possibilities of intensive biodynamic farming, had laid the foundations for more decentralized urban development in small units, responsive to direct human contact, and enjoying urban and rural advantages.[22]

If Kropotkin felt that Morris was too naïve and anti-urban, Errico Malatesta criticised Kropotkin in turn for his belief that the urban general strike would usher in revolutionary change. Malatesta was a realist; he argued that (and examples in the twentieth century bear him out) modern cities relied on just-in-time provisioning, so the city would starve during a general strike and the forces of the state could wait out the revolt.[23] Besides, Malatesta also insisted that continuity was key: the power plants needed to remain in operation, the city needed its provisioning agents in the countryside and the networked life of urban industrial society needed to be maintained. So during the occupation of the factories in Italy in 1920, he pleaded with the occupiers of a large plant in Milan to restart production and exchange or face the consequences of a backlash, which of course was the rapid rise of Fascism.[24] As we shall see, later similar challenges were encountered by the anarchists of Barcelona when power was handed to them in the summer of 1936. But this discussion of realism returns us to the grim realities of the how the Commune of Paris was terminated.

After vicious fighting, governmental forces carried out mass executions of men, women and children with anti-Communard civilians joining in the massacre. John Merriman estimates more than 100,000 residents of Paris were killed, imprisoned or fled the scene: some 15,000 were shot out of hand in the weeks following the suppression of the Commune. While the skilled trades: shoemakers, tailors, cabinetmakers, bronze workers, plumbers (all trades that were the backbone of the Commune,) were so decimated that 'industrialists and small employers complained about the paucity of artisans and skilled workers'.[25] The Communards had executed under a hundred hostages. For the 'men of order', Merriman reports, the entire population of Paris was guilty. One 'man of order' dreamed of 'an immense furnace in which we will cook each of them [the citizens of Paris, CL] in turn'.[26] And thus as Ross concludes, 'the attempt on the part of the bourgeois-republican government to physically exterminate its class enemy bears every resemblance to mass exterminations of religion and race',[27] unleashing, Merriman adds, 'the demons of the twentieth century'.[28] The symbolism of the subversive heights of the Buttes-Chaumont was understood by the 'men of order', who ordered the construction of Sacré Coeur where the National Guards' artillery had been parked.[29]

The Paris Commune served as a catalysing agent in which the anarchist and the libertarian wings of the First International coalesced (an example of 'globalisation from below'). In Italy and Spain, the example of decentralised federations of cities fell on fertile ground. In Italy radical Mazzinians disavowed Mazzini when he denounced the Commune as a breeder of class war and godlessness, and they turned their backs on his centralising precepts. In Spain, the most radical communal federalist Republicans were one of the streams from which the Spanish anarchist movement developed, especially after Spain's own short-lived Commune at Alcoy in 1873.[30]

Several years later, as the direct action and insurrectional techniques of the anti-authoritarian branch of the First International ran into an impasse, the politics of the city and the urban commune were the route used to escape this ineffective radicalism. So one path to gas and water socialism on the continent was promoted by the former anarchist firebrands, such as the French Paul Brousse[31] and the Italian, Andrea Costa,[32] or the more moderate Communard exile based in Palermo and then Milan, Benoît Malon,[33] who championed a city based experimentalist socialism.[34] The communal experiment was invoked as a reformist but radical programme at the city level, indeed in Costa's case, maximalist socialism mixed with an older tradition of *campanilismo*. Thus for his followers in the restive Romagna, expanding male suffrage made Bologna, Ferrara and smaller cities targets from which to control the growing armies of landless labourers in the adjacent commercial farming lands of the Po Valley. These cities and their rural hinterlands became the backbones of Italian socialism before they were smashed by Fascism in the early 1920s. The Socialists in 1920 mixed a minimum programme of radical gas and water socialism with a maximalist programme, which envisaged the socialisation of the land and the creation of urban soviets. The political geography of Padania, forged in the late nineteenth century, in the aftermath of the First International, was transformed by the Communists after 1945 into the prosperous but left-wing Red Belt.[35]

Within the broader boundaries of the Red Belt, the anarchists developed generational fortresses: to name four, Ancona, Massa-Carrara, Livorno and La Spezia. This network retained anarchism's presence in Italy from the 1890s to the 1920s, even as the parliamentary socialists of the Partito Socialista Italiano became dominant on the Left, with Costa as one of its early leaders in parliament a dominant player on the Left.[36]

In June 1914, on the eve of the First World War, the power of this more radical network of small to medium-sized towns was vividly demonstrated during the so-called Red Week. Like the Tragic Week in Barcelona in 1909, the combination of anti-militarism caused by unpopular imperial adventures and the miserable treatment of conscripts, rising inflation, especially for basic food stuffs, and deeply embedded anti-clerical and republican sympathies, led to general strikes, police shootings and general uprisings, in Barcelona in 1909 and in 1914 throughout the web of towns in central Italy, which saw the peninsula nearly cut in two, the declaration of republics in Romagnole towns and the raising of Trees of Liberty (the Great French Revolution was still a living political tradition here).[37] Max Nettlau, the 'Herodotus of Anarchism', reported, when the events were still unfolding, that the small towns of central Italy had retained their revolutionary spirit whereas in Milan, Turin or Genoa events were less dramatic.[38] Nettlau's anarchism was suspicious of large organisations but the spirit he identified in the Romagnole towns could also be found in the distinctive areas of great cities, such as Paris's Belleville or on the Buttes-Chaumont, or in the fast growing but isolated industrial suburb of Borgo San Paolo in Turin during the First World War, which was one of the hearts of the urban insurrection in 1917, partially inspired by events in Dublin a year earlier, and witnessed a war-weary, bread-hungry populace charged by events in Russia and local anarchists, march on the bourgeois centre of Italy's 'motor city'.[39]

If Costa's usage of the Paris Commune was an uneasy mixture of reformism and revolutionary rhetoric, the former anarchist Paul Brousse and the Belgian libertarian social democrat, Cesar de Paepe, chose to advance a model which they felt avoided the bureaucratic and centralising tendencies of the German Social Democratic party but mixed the libertarianism of the anarchist tradition with electoral forms of social democracy. Socialism at the municipal level would be more democratic and efficient. Thus de Paepe argued for the

democratic control of local utilities, municipal bakeries and public, cooperative housing. His model avoided statist and bureaucratic dangers because the
local citizenry and the workers of the utilities would manage and control the
sinews of the infrastructure from which a socialist society would be created.
Local democracy at the borough, commune or parish level and worker's
control in the municipal utilities would avoid anonymous statist ownership
(thus a communalist rather than an anarchist or social democrat path to the
future city).[40]

This model of decentralised democratic gas and water socialism was dis-
seminated throughout Europe and influenced the Fabians in Britain. Of
course, when one hears the words Fabian or Fabianism, the images which
come to mind are of bureaucratic control by experts, or the lavish praise of
the Webbs for the High Stalinist Soviet Union's new civilisation or Bernard
Shaw's appreciation of Mussolini's social programmes in the 1920s. This is
misleading. Early Fabian history is quite different. Charlotte Wilson, along
with the Italian Francesco Merlino, championed an anarchist wing of the
Fabians in the 1880s and in the early twentieth century G.D.H. Cole championed guild socialism.[41] In the 1890s the Webbs took anarchism or small 'a'
anarchism very seriously indeed: their massive studies of the history of trade
unionism and industrial democracy were driven by a need to show how
'primitive' direct democracy in workers' organisations was outlasting its usefulness but was certainly not ignored and their next great project was a study
of local government in England.[42] Indeed, in the 1880s both Beatrice Potter
and George Bernard Shaw had been attracted to individualist anarchism and
part of their conversion to Fabianism involved journeying down convoluted
roads away from this attraction. In Bernard Shaw's case it was a mutualist
Proudhonian Communard refugee who introduced him to socialism in the
first place. 'We had not sorted ourselves out', he recalled.[43] So the gas and
water socialism of the Fabians and the Webbs's appreciation for local government must be placed in a larger zeitgeist, which also included the lessons
and influences of the Paris Commune.

Cities, anarchism and the global south in the late nineteenth and early twentieth century

I now return to a theme flagged at the beginning of this chapter, classical
anarchism and the Global South. The city has been central to recent pioneering
research and has been carried out by piecing together the transnational networks, which bound the world together in this first era of modern globalisation. The spread of anarchism and syndicalism followed the circuits of power,
the circuits of migration, the paper, print and human links of diasporic communities, of language communities and within cosmopolitan melting pots of
anarchist and syndicalist conviviality and politics.[44] How does one trace these
circuits? Simply, by identifying port cities and global hub cities in which one
can identify longstanding or temporary communities of native, immigrant

and nomadic anarchists and syndicalists. There are two ideal-typical networks: those of political refugees, another which mixes that status with individuals who follow the circuits of imperialism and capitalism, which is composed of labourers, skilled workers and sailors. Both carry anarchism and syndicalism into new environments and pitch up in a series of large and medium-sized ports. Thus, for example, there is the back and forth between Argentina and Spain, which can be traced for fifty years between Buenos Aires and the vast hinterland of the Rio de la Plata (with its engrained port-based syndicalist political culture spreading deep into the interior) and Spanish and Catalan port cities.[45] Thus repression and search for work drove Spanish anarchists to Argentina in the 1890s and 1920s and after a military coup in Argentina in 1931 anarchists and syndicalists returned to the Spanish Republic and played a signal role in the Civil War, only once again after their defeat to flee largely to Mexico and South America.[46]

The Industrial Workers of the World (IWW or 'Wobblies') organised sailors, stevedores and oil workers of the West (and East) Coast of the USA, Mexico, Peru and Chile; and those Wobbly locals stretched down the coast from San Francisco and San Diego to Baja California towns, Peruvian port towns and Valparaiso in Chile. In turn these efforts spread inland so that Santiago's vibrant anarchist scene was seeded by those nomadic Wobblies, whereas in Peru the anarchists and syndicalists in coastal towns linked up with indigenous peoples and organised with them against a semi-feudal system of state-enforced corvée labour.[47] The anarchist inspired Liberal Party of Mexico used a base in San Diego to organise an invasion of Baja California during the Mexican Revolution.[48] And a recent study shows how Los Angeles and San Diego became melting pots of Anglo, immigrant and Mexican labourers and radicals, involved on both sides of the border.[49] In East Asia, anarchist and syndicalist students and intellectuals, used Shanghai, Canton, Hong Kong and Tokyo as centres of refuge, study and plotting. Thus, for example, Vietnamese radicals were inspired by Chinese and Korean anarchists in Shanghai and Hong Kong.[50]

The modernisation of Egypt and the rise of the cotton cash-crop, factories and the building of the Suez Canal attracted peasants from the Egyptian countryside but also workers and artisans from the Ottoman Empire, Greece, Italy, the Habsburg lands and Czarist Russia. From these parts anarchists and syndicalists arrived in Alexandria and Cairo, and as Ilham Khuri-Makdisi shows in her comparison of Beirut, Cairo and Alexander, a variety of secular radical movements thrived, which owed a great deal to anarchism and the related rationalist educational theories and practices of Francisco Ferrer. Although in these cities, the European quarters and indigenous Christians were more likely to be attracted to these movements, solidarity cemented by struggles against entrepôt capitalists dissolved some of the sectarian boundaries between Muslims and non-Muslims.[51] Other studies have traced hubs such as Havana, from where anarchists and syndicalists spread their ideas and practices, at Tampico in Mexico, Ybor City in Florida, the Panama Canal

Zone with its vast labouring force, and San Juan in Puerto Rico.[52] The Italian diaspora is an extraordinary example of city networks spanning the globe and feeding back to the Italian movement in the peninsula. By utilising newspaper subscriptions, Davide Turcato has mapped out networks from Alexandria, to Paterson, New Jersey and New York City to São Paolo and Buenos Aires, while more recent studies have used similar techniques to map city and town networks covering Canada and the USA. I have done the same thing in following the movements of Errico Malatesta, who made London home for nearly thirty years.[53] Whereas Kenyon Zimmer's study of cosmopolitan San Francisco is a case apart. Here no single immigrant or native anarchist group dominated, thus East European Jewish and Italian anarchists arrived from the East Coast, Mexican radicals from the south and Chinese and Japanese anarchists from the East joined by the strong presence of Indian anti-colonial radicals of the Ghadar movement, who at this point were attracted to anarchism. Cowboys, hoboes, former hard rock miners and bindlestiffs from Irish and Anglo-Saxon backgrounds pitched up too, and thus activities were pointed to homelands, city politics, the Mexican Revolution and the vigorous organization of farm labourers by the IWW in the extraordinarily fertile Central Valley of California.[54]

One might also trace the movements of intellectuals and professionals from the Global South to imperial cities, just as had been done previously for anticolonial nationalist elite formation and indeed this fashion for tracing networks of anarchists began with that premier student of nationalism, the late Benedict Anderson, in his study of the life of José Rizal, novelist, anarchist and national martyr of the Philippines, who journeyed from Manila to Hong Kong, to Barcelona, Paris, various German cities and London. Here the mixing of Tagalog, Spanish, French, German and British cultures unveils a fascinating life story.[55] Thus, starting with Anderson's work and my forays into the life and times of Errico Malatesta, this field has expanded into a series of cartographies of anarchism and radicalism through network analysis of urban hubs and port cities, and thereby presents us with an alternative modernity, an alternative globalisation, during the era of High Imperialism.

London was the capital of the capitalist world and the centre of the greatest empire on Earth, and the host to interconnecting anarchist colonies originally based in Soho and the East End but gradually suburbanised by the spreading of the Underground Railway. The city was populated by exiles from the continent, even liberal Switzerland, who had fled or had been deported to the sole remaining country in Europe whose asylum laws were more flexible and open than her neighbours', although the era of dynamite and assassinations in the 1890s and the 1905 Aliens' Act tightened up on flows of anarchists and so-called paupers. However there were also limits to the cosmopolitan nature of these colonies, since language communities could limit intermingling and could lead to tensions, equally exacerbated by the spies and police agents of a dozen interested foreign governments. These of course found an echo in fictional accounts of this 'exile' and 'immigrant' London, most famously by another immigrant, the Polish novelist and merchant sea captain denizen of the globalisation of High Imperialism, Joseph Conrad.[56]

Art, anarchism and the city

From the 1880s to the present, the artists, the art market, the anarchist and the urban bohemia have been complex constant features of the city. From Pissarro to Rothko and beyond, certain districts (Montmartre or Greenwich Village, Lower Eastside, etc.) hosted an urban ecology characterised by cheap rents, immigrants, the marginalised or the peripheral working classes, and a network of cheap restaurants, cafés, cabarets, dance centres, radical churches, unorthodox bookshops, the list could go on. As myriad studies of the Paris-based impressionists, cubists and early surrealists show, the political economy of Montmartre, anarchist argot and the humour of cabaret performers, were interlaced with the aura of daring surrounding the artists' work. There was a need for an urban art market in cities with national and international pre-dominance (such as Paris or New York) in which bourgeois critics such as the mercurial Felix Fénéon (civil servant, high bourgeois, art critic journalist and possible terrorist in the 1890s) acted as mediators between bourgeois society and this milieu. These liminal actors served as mediators between bourgeois society and this milieu: pathfinders, arbiters and patrons of the new schools and art markets. Thus the commodification of the daring and the commodification of the rebels were interlaced with new markets for capitalist penetration, anticipating themes that have a contemporary ring to them.[57]

In general the relationship between the artistic and literary worlds and anarchism is a complex one. Bohemia and the world of the dandy were originally apolitical or indeed right-wing and elitist, but it is the case that a certain reading of Nietzsche, Ibsen, Kropotkin or syndicalist ideas might turn writers and artists to anarchism for at least a period of their lives: O'Neill, Joyce, Kafka, the list is rather long.[58] But one should be careful in drawing clear lines of influence, because the effect of anarchism on their work ranged from the muse and provocateur, to technical guide to their literary or painterly style, to a deep and long-lasting commitment to the movement.[59] Thus, for example, Pissarro and Signac were committed to the movement whereas Picasso ingested a certain energy from the anarchists of Barcelona.[60] Previously, I mentioned how global ports and hubs acted as transmission and ideational houses of exchanges in a global network of anarchist and syndicalist organisers and militants, so too the global network of libertarian artistic bohemia can be traced through the peregrinations by mobile artists (Man Ray, for example, from a New York to a Parisian setting), but also by mobile self-educated activists bridging the world of art and literature with the more focussed worlds of anarchist politics and anarchist culture.[61]

Perhaps the best example of the latter is Emma Goldman. To understand Goldman one should invoke a popular term in radical intellectual and activist circles of today: intersectionality.[62] Thus according to the concept of intersectionality, or the matrices of domination, power relations (capitalism, racism, patriarchy, age etc.) overlap and exert mutual influences on each other, with individuals and institutions being placed at the intersections of these systems of domination. Goldman suffered from an abusive father in Czarist Russia; she was the victim of abusive relationships from Johann Most to Ben Reitman, but she also had been a sweatshop worker in capitalist America; she was a Jew in a world of pogroms and widespread anti-Semitism; she was an immigrant from the suspect not-quite-white extremities of East Europe; she was a feisty self-educated radical in the schizophrenic world of plutocratic American democracy: intersectionality explains her biography.[63]

Goldman became a leading conveyor of the avant-garde of Greenwich Village and Provincetown to the timid if titillated WASP [White Anglo-Saxon, Protestant] provincial middle classes. One could map her biography of intersectionality onto the social geography of Gilded and Progressive Age New York. Thus from the nerve centre of her journal *Mother Earth*, first in Greenwich Village and then in Harlem, a network of connections arose.[64] *Mother Earth* was in many respects a rather outré member of those little magazines, which shook the cobwebs from staid WASP literature and art, albeit Margaret Anderson still exhibited the sniffy attitude of the 'original' Americans about these rather odd, rather shakily educated, and threatening but alluring interlopers. Goldman's 'Pilgrim's Progress' from sweatshop worker in Rochester, New York to the liberating possibilities of Manhattan is mapped out in this geography.[65] Goldman's introduction to politics and literature came largely through an earlier Lower Eastside network of anarchists, the same network described in *Beer and Revolution*, but filtered through a culturally secular but Jewish milieu of cafés and restaurants, in which Yiddish and Russian, not German, were the shared lingua franca. The all-American, Anglophone Red Emma of the immediate years before 1914 came later, where in small-town America, at the gatherings which combined politics with the open air entertainment of the tent camp evangelist, she proclaimed the right of women to control their own bodies but also the ennobling virtues of Whitman, O'Neill, Nietzsche, Hauptmann and Ibsen.

In any case, she was in her pomp in the tiny offices of *Mother Earth*. In a richly radical Manhattan geography where the Ferrer School, radical Greenwich Village churches, branches of the IWW and her journal mixed, for a brief moment (1914), as Thai Jones shows in his superb portrait of this world, anarchists, anarchist bohemians, Wobblies and unemployment marches gripped the city's imagination.[66] A few years earlier, the Wobbly-led strike at the nearby silk mills of Paterson, New Jersey, the host to a lively community of Italian anarchists mentioned earlier, gave force to the 'Paterson Pageant' in Manhattan, organized with the help of the Western 'exotic' of the hard rock miners, Big Bill Haywood, with the assistance of Emma Goldman and John Reed (memorably depicted in Warren Beatty's film *Reds*), which occurred almost concurrently with the revolutionary Armoury Show of Post-Impressionist/Cubist paintings.[67] And Big Bill was not immune to the attractions of Greenwich

Village bohemia, with other comrade Wobblies complaining that he had gone soft and could be found writing poetry on the park benches of Washington Square.[68] The anarchism of Emma Goldman and Alexander Berkman (lover, comrade and lifelong friend) was inseparable from the physical presence of New York. Berkman lamented as the ship deporting them to revolutionary Russia passed through Lower New York Bay in the early morning.[69]

> Slowly the big city receded, wrapped in a milky veil. The tall skyscrapers, their outlines dimmed, looked like fairy castles lit by the winking of stars and then all was swallowed in the distance.

The Futurists in Milan before 1914 were closely associated with anarchism. Here, too, we have another example of the anarchist imagination encountering the urban, modernist frisson of the pulsating city.[70] The first painting considered Futurist, Carlo Carrà's 'The Funeral of the Anarchist Galli', depicts a ferocious and confused flight of anarchist mourners, charged by the police in the streets of Milan. The inspiration was far from the bucolic dreams of William Morris's garden Trafalgar Square or even the fields, factories and workshops and garden cities of Kropotkin or Geddes. The city was speed, confusion, danger, bricks and mortar, sensation and lights. Thus Carrà describes the genesis of the drawing and subsequent painting:

> I found myself unwillingly in the centre of it, before me I saw the coffin in red carnations sway dangerously on the shoulders of the pallbearers; I saw horses go mad, sticks and lances clash, it seemed to me that the corpse could have fallen to the ground at any moment and the horses would have trampled it. Deeply struck, as soon as I got home I did a drawing of what I had seen.[71]

However, there were urban settings where bohemia briefly impacted significantly on politics and State power. Under the influence of Dada in Zurich in 1917–1918,[72] the so-called writer's and poets' revolution of the Munich Soviet in 1919 is perhaps the most famous.[73] But as Roy Foster shows, a mixture of Irish Republican politicos, syndicalists and Dublin's avant-garde seized the Post Office in 1916.[74] In Munich in 1919, anarchists and their artistic followers ran a Soviet for one week, only to be displaced by the far more ruthless Jacobin Bolsheviks, only in turn for them to be eliminated by the proto-Fascist Freikorps. The image of urban Cultural Bolshevism, as the Nazis deemed it, of Schwabing armed (with Corporal Hitler cooling his heels in the barracks during the Soviet episode), led incongruously by the writer and anarchist pacifist Gustav Landauer, the Expressionist playwright, invalided war veteran Ernst Toller and the author Erich Mühsam, and others, became an abiding theme in Nazi propaganda throughout the 1920s and 1930s. Many of these revolutionaries shared with others on the Right, the life reform movements, the cults of sun and dance, monetary quackery and the Laban dance experiments, but the war and the Munich Soviet served as a bloody, unbridgeable chasm.

In a similar fashion, but with different lead players, bathed in a cult of violence, the poet D'Annunzio and his Legionaries seized the city of Fiume on the Adriatic coast in 1919 and claimed it for Italy. A correspondent for the Milanese anarchist daily, *Umanità nova*, noted several former comrades from the 1914 Red Week in his ranks. The city of Fiume became a theatre for D'Annunzio's experiment in ultra-nationalism, but an ultra-nationalism also connected to a bohemian, life reform and drug addled *zeitgeist*. Indeed for a moment there were suggestions of a March on Rome, to overthrow the hapless post-war government, in which the Legionaries, the Maximalist Socialists, and the briefly very popular anarchists, led by the charismatic Errico Malatesta, would all participate. While Mussolini absented himself from this March on Rome, it was precisely D'Annunzio's *Arditi* shock troops (once the government dispersed the grotesque theatre at Fiume, with a taste of grapeshot at Christmas of 1920) who would launch the Fascist counter-revolution and deliver havoc in those Red Belt towns of Padania, previously mentioned in reference to the Red Week of 1914.[75]

Red and Black Barcelona in 1936: Paris Commune Redux

In the summer of 1936 the CNT–FAI [Confederación Nacional del Trabajo– Federación Anarquista Ibérica], the anarcho-syndicalist movement of Spain were the masters of one of the great modern industrial, commercial and intellectual cities of Europe. Just as the Paris Commune of 1871 loomed large in the memories and imaginations of Marxists and Socialists in the half century after its bloody suppression, so too have the multiple images of Red and Black Barcelona generated heat and some light within the Left and the ex-Left ever since.

Anarchism gained a grip on Barcelona for two reasons. First, the anarcho-syndicalist union, the CNT, grew to an immense size during the rapid industrialisation of the city as a centre of war production in neutral Spain during the First World War. Second, we must point to the growth of Barcelona's suburbs, where migrants streamed into the building trades as the continuous growth of bourgeois, art deco, Barcelona demanded more and more skilled and unskilled labour. In the suburbs, with their jerry-built housing, unscrupulous landlords, lack of well-established Catholic clerical networks, the anarchists took on the mantle of community organisers (rent strikes etc.) and political recruiters, and established street by street strongholds: thus industrial strategies and life in the city gave the anarchists remarkable continuity, even if their ranks were thinned by employer gunmen in the early 1920s and the suppression by the dictatorship Primo de Rivera for the rest of that decade. But there was also a growing linkage with the left-wing of middle class Catalan nationalism, and indeed Barcelona-born reformists in the CNT sought common ground.[76]

But once the anarchist militias were triumphant in the summer of 1936, the anarchists engaged in workers' self-management of large and medium sized firms, they instituted new forms of education, they sought a more comprehensive health system and to regulate drinking and sex workers, they sought neighbourhood forums of democracy, they used large luxurious hotels as popular restaurants and sought to make visual art, theatre and music more accessible to the general public. They even engaged in their own version of gas and water socialism, through the control of the municipal water company.[77]

But the full anarchist package was not on offer. Due to the threat from Franco's forces, Hitler and Mussolini, and the need by the Republican forces for outside support (increasingly the Soviet Union and in consequence within the besieged Spanish Republic, the growing power of Spanish Communism), some of the more radical figurers of pre-1936 anarchism joined the national government. But even before the Soviet Union and the Spanish Communists were a formidable force on the domestic scene, the anarchists of Barcelona refused to take power when handed it by the President of Cataluña and instead established an anti-fascist coalition to run urban affairs. On 21 July 1936, municipal committees were established for supplies, transportation and production, but when the CNT met on the 23 July, the motion to institute libertarian communism in Barcelona was defeated: anarchists did not seize power like Jacobins or Bolsheviks, anarchist speakers insisted, and instead they joined a cross-party anti-fascist coalition which in turn formed a Catalan government.[78] The war prevented further radical changes and the imperatives of war production increased the hierarchy within the industrial economy: shortages and rationing brought internal tensions and the controversy concerning the militarisation of the militias was married to suspicions about the growth of the Communists and their direct and indirect influence on the formidable Socialists. The crisis came to a head over a conflict about regaining governmental control from anarchist elements of the central telephone exchange, and street fighting ensued in the heart of Barcelona in May 1937. In the end the anarchist leadership gave in because they would not call their armed supporters back from the front and endanger the war effort. Hundreds died in street fighting, and in the aftermath Soviet Union-influenced elements of the security forces hunted down the irreconcilable anarchists but more pointedly the dissident Marxists of the POUM [Partido Obrero de Unificación Marxista], including that eyewitness to Red and Black Barcelona, George Orwell.[79] And it is Orwell's vivid description of the only city he had been in where the working class 'was in the saddle' and even the bootblacks had been collectivised and had painted their shoe boxes red and black, and the waiters looked you in the face, which still lingers in our imaginations.[80]

One shouldn't forget the darker sides of the scene: the destruction of churches, spurred on by the intense loathing of the Catholic hierarchy by much of the popular classes and at least in the summer of 1936, the ride out of town and the bullet in the head for identifiable enemies, albeit according to Paul Preston's accounting, the forces of Franco carried out far more executions in their areas.[81]

The entry of Franco's troops into Barcelona in 1939 has been used to signal the end of Classical Anarchism, whether or not the new anarchism after 1945 may have had more continuities with the past than has been assumed I will leave unanswered, but what is clear is that encounters of anarchism with the city still remained and remains a vital optic to understand how anarchism was transformed from a philosophy to an ideology and praxis.[82]

Notes

1 Some of this chapter formed part of my Professorial Inaugural Lecture held at Goldsmiths, University of London (9 June 2016).
2 For the period after 1945 see, Levy, 'New Anarchism and the City from the Cold War to the Occupy and Square Movements'.
3 For a general overview see, Levy, 'Social Histories of Anarchism', 10–15.
4 Levy (ed.), *Colin Ward. Life, Times and Thought*.
5 Konishi, *Anarchist Modernity: Cooperation and Japanese-Russian Intellectual Relations in Modern Japan*; Ramnath, *Decolonizing Anarchism: An Antiauthoritarian History of India's Liberation Struggle*.
6 Maxwell and Craib (eds), *No Gods No Masters No Peripheries. Global Anarchisms*.
7 Hirsch and van der Walt (eds), *Anarchism and Syndicalism in the Colonial and Postcolonial World, 1870–1940* [hereafter Hirsch and van der Walt, *Anarchism and Syndicalism*].
8 Bantman and Altena (eds), *Reassessing the Transnational Turn: Scales of Analysis in Anarchist and Syndicalist Studies* [hereafter Bantman and Altena, *Reassessing*].
9 Goyens, *Beer and Revolution: The German Anarchist Movement in New York City, 1880–1914*, 17–18; in general see Goyens, 'Social Space and the Practice of Anarchist History', *Rethinking History*, 439–457; and see the much anticipated, Goyens (ed.), *Radical Gotham: Anarchism in New York City from Schwab's Saloon to Occupy Wall Street*.
10 Di Paola, *The Knights Errant of Anarchy. London and the Italian Anarchist Diaspora (1880–1917)* [hereafter Di Paola, *Knights Errant*].
11 Bantman, *The French Anarchists in London, 1880–1914: Exile and Transnationalism in the First Globalization* [hereafter Bantman, *French Anarchists*].
12 Dahlmann, *Land und Freiheit: Machnovščina und Zapatismo als Beispiele agrarrevolutionärer Bewegungen*; Shubin, *Nestor Machno: bandiera nera sull'Urcaina. Guerriglia libertaria e rivoluzione contadina (1917–1921)*.
13 Scott, *The Art of Not Being Governed: An Anarchist History of Upland Southeast Asia*.
14 Linebaugh and Rediker, *The Many Headed Hydra*.
15 Ross, *Communal Luxury. The Political Imaginary of the Paris Commune* [hereafter Ross, *Communal*], 44–45.
16 Ross, *Communal*, 67–142.
17 Malatesta, 'La Comune di Parigi e gli anarchici' and similar critique by a Communard, Élisée Reclus, Fleming, *The Geography of Freedom. The Odyssey of Élisée Reclus*, 89–92 and Ferretti, *Anarchici ed editori. Reti scientifiche, editoriali e lotte culturali attorno all Nuova Geografia Universale di Élisée Reclus (1876–1896)*, 47–49.
18 Ross, *Communal*, 104–116.
19 Ross, *Communal*, 60.
20 Kinna, 'Kropotkin's Theory of the State: A Transnational Approach', in Bantman and Altena, *Reassessing*, 43–61 and for a critical account of the urban guilds see

22   *Carl Levy*

MacLaughlin, *Kropotkin and the Anarchist Intellectual Tradition*, 229–234. Also
see Sennett, *Flesh and Stone: The Body and City in Western Civilization*, 200–202.
21 Ryley, *Making Another World is Possible. Anarchism, Anti-Capitalism and Ecology
in Late 19th and Early 20th Century Britain* [hereafter Ryley, *Making*], 155–188.
22 Mumford, *The City in History*, 585.
23 Malatesta, '*Lo Sciopero Armato*': Il lungo esilio londinese, 1900–1913.
24 Malatesta, 'La propaganda del compagno E. Malatesta'.
25 Merriman, *Massacre. The Life and Death of the Paris Commune* [hereafter Merriman, *Massacre*], 253–256.
26 Merriman, *Massacre*, 255.
27 Ross, *Communal*, 81.
28 Merriman, *Massacre*, 256.
29 Mitchell, *Cultural Geography*, 112.
30 Civolani, *L'anarchismo dopo La Comune. I casi italiano e spagnolo*; Esenwein,
Anarchist Ideology and the Working-Class Movement in Spain, 1868–1898, 45–48.
31 Stafford, *From Anarchism to Reformism: A Study of the Political Activities of Paul
Brousse with the First International and the French Socialist Movement, 1870–1890*.
32 Levy, 'Italian Anarchism, 1870–1926' [hereafter Levy, 'Italian'] in Goodway (ed.),
For Anarchism, History, Theory, and Practice [hereafter Goodway, *For Anarchism*], 38; Di Maria (ed.), *Andrea Costa e il governo della città. L'esperienza
amministrativa di Imola e il municipalismo popolare (1881–1914)*.
33 Vincent, *Between Marxism and Anarchism: Benoît Malon and French Reformist
Socialism* [hereafter Vincent, *Between*].
34 Dogliani, *Un laboratorio di socialismo municipalismo. La Francia (1870–1920)*.
35 Crainz, *Padania: Il mondo dei braccianti dall'Ottocento all fuga dale campagne*.
36 Levy, 'Italian', 34.
37 Levy, 'Italian', 54–56.
38 Max Nettlau (Vienna) to Thomas Keell (London), 22 June 1914 (IISH [International Institute of Social History, Amsterdam]: Freedom Press Archives).
39 Levy, *Gramsci and the Anarchists*, 89–94.
40 Vincent, *Between*, 151; Gaspari and Dogliani, (eds), *L'Europa dei comuni: Origini
e sviluppo del movimento communale europeo dall fine dell'Ottocento al secondo
dopoguerra*.
41 Ryley, *Making*, 59, 70, 90, 103, 119, 132; and Stears, 'Guild Socialism', in M. Bevir
(ed.), *Modern Pluralism. Anglo-American Debates since 1880*, 40–59.
42 Harrison, 'Sydney and Beatrice Webb', in C. Levy (ed.), *Socialism and the
Intelligentsia 1880–1914*, 56–64.
43 Cited in Ross, *Communal*, 111.
44 Levy, 'Anarchism and Cosmopolitanism', 256–278.
45 De Laforcade, 'Straddling the Nation and the Working World: Anarchism and
Syndicalism on the Docks and Rivers of Argentina', in Hirsch and van der Walt,
Anarchism and Syndicalism.
46 Baer, *Anarchist Immigrants in Spain and Argentina, Anarchism and Syndicalism*,
321–362.
47 Hirsch, 'Peruvian Anarcho-Syndicalism: Adapting Transnational Influences and
Forging Counterhegemonic Practices', in Hirsch and van der Walt, *Anarchism and
Syndicalism*, 227–272; Rosenthal, '"Moving between the Global and the Local".
The Industrial Workers of the World and their Press in Latin America', in de
Laforcade and Shaffer (eds), *In Defiance of Boundaries. Anarchism in Latin American
History*, 72–94 [hereafter de Laforcade and Shaffer, *In Defiance*]; Craib, *The Cry of
the Renegade: Politics and Poetry in Interwar Chile*.
48 Sandos, *Rebellion in the Borderlands: Anarchism and the Plan of San Diego, 1860–1923*.
49 Struthers, '"The Boss has no Color Line": Race, Solidarity, and a Culture of
Affinity in Los Angeles and the Borderlands, 1907–1915', 61–92.

50 Hwang, 'Korean Anarchism before 1945: A Regional and Transnational Approach', in Hirsch and van der Walt, *Anarchism and Syndicalism*, 95–129.

51 Khuri-Makdisi, *The Eastern Mediterranean and the Making of Global Radicalism, 1860–1914*.

52 Shaffer, 'Tropical Libertarians: Anarchist Movements and Networks in the Caribbean, Southern United States, and Mexico, 1890s–1920s', in Hirsch and van der Walt, *Anarchism and Syndicalism*, 273–320; Shaffer, 'Panama Red: Anarchist Politics and Transnational Networks in the Panama Canal Zone, 1904–1913', in de Laforcade and Shaffer, *In Defiance*, 48–71.

53 Turcato, 'Italian Anarchism as a Transnational Movement 1885–1915', *International Review of Social History*, 407–444; Tomchuk, *Transnational Radicals, Italian Anarchists in Canada and the U.S. 1915–1940*; Levy, 'The Rooted Cosmopolitan: Errico Malatesta, Syndicalism, Transnationalism and the International Labour Movement', in Berry and Bantman (eds), *New Perspectives on Anarchism, Labour & Syndicalism: The Individual, the National and the Transnational*, 61–79.

54 Zimmer, 'A Golden Gate of Anarchy: Local and Transnational Dimensions of Anarchism in San Francisco, 1880s–1930s', in Bantman and Altena, *Reassessing*, 100–117 and for urban Yiddish and Italian anarchists in the US, see Zimmer, *Immigrants against the State. Yiddish and Italian Anarchism in America*.

55 Anderson, *Under Three Flags. Anarchism and the Anti-Colonial Imagination*.

56 Di Paola, *Knights Errant* and Bantman, *French Anarchists*. For Conrad and London as the city of world imperialism and anarchism, see Ó Donghaile, *Blasted Literature. Victorian Political Fiction and the Shock of Modernism*, 113–129.

57 Halperin, *Felix Fénéon: Aesthete and Anarchist in Fin-de-siècle France* [hereafter Halperin, *Felix Fénéon*]; Sonn, *Anarchism and Cultural Politics in Fin-de-siècle France*; Brigstocke, *The Life of the City: Space, Humour, and the Experience of Truth in Fin-de-siècle Montmartre*; Mould, *Urban Subversion and the Creative City*. Also how the confirmed anarchist, Camille Pissarro reflected on the urban art market, Brettell, *Pissarro's People* [hereafter Brettell, *Pissarro's*], 219–239.

58 Gluck, *Popular Bohemia: Modernism and Urban Culture in Nineteenth-Century Paris*; Shantz, *Against all Authority. Anarchism and the Literary Imagination*; Shantz (ed.), *Specters of Anarchy. Literature and the Anarchist Imagination*.

59 For a magisterial overview of the theme and also of the class tendencies with the anarchist literary and artistic intelligentsia see Cohn, *Underground Passages. Anarchist Resistance Culture 1848–2011*. For painting see Naumann, 'Aesthetic Anarchy', in Jennifer Mundy (ed.), *Duchamp, Man Ray Picabia*, 58–76 and Leighton, *The Liberation of Painting. Modernism and Anarchism in Avant-Guerre Paris*.

60 Leighten, *Re-ordering the Universe: Picasso and Anarchism*; Kaplan, *Red City, Blue Period. Social Movements in Picasso's Barcelona*; Halperin, *Felix Fénéon*; Varias, *Paris and the Anarchists: Aesthetes and Subversives during the Fin-de-Siècle*; Brettell, *Pissarro's*.

61 Antliff, *Anarchist Modernism: Art, Politics and the First American Avant Garde*.

62 Dupuis-Déri, 'Is the State Part of the Matrix of Domination and Intersectionality? An Anarchist Inquiry', 13–35.

63 Ferguson, *Emma Goldman. Political Thinking in the Streets*, 67–175.

64 Glassgold (ed.), *Anarchy! An Anthology of Emma Goldman's Mother Earth* and Stansell, *American Moderns. Bohemian New York and the Creation of a New Century*, 132–134, 137, 200 [hereafter Stansell, *American*].

65 Stansell, *American*, 114, 202–206.

66 Jones, *More Powerful than Dynamite. Radicals, Plutocrats, Progressives and New York's Year of Anarchy*.

67 For the connections to the Italian anarchists through the Italian anarchist journalist Carlo Tresca, see Pernicone, *Carlo Tresca: Portrait of a Rebel*, 61–73; *Reds*, Director Warren Beatty, Paramount Pictures, 1981.

68 Stansell, *American,* 151.
69 Fellner (ed.), *Life of an Anarchist. The Alexander Berkman Reader,* 163.
70 Berghaus, *Futurism and Politics: Between Anarchist Rebellion and Fascist Reaction, 1900–1944.*
71 Carrà, *La Mia Vita,* 50; and see C. Poggi, *Inventing Futurism. The Art and Politics of Artificial Optimism.*
72 Brendell. 'The Growing Charm of Dada'.
73 Levy, 'Max Weber, Anarchism and Libertarian Culture: Personality and Power Politics', in Whimster (ed.), *Max Weber and the Culture of Anarchy,* 83–109, and Landauer, *Revolution and Other Writings: A Political Reader.*
74 Foster, *Vivid Faces. The Revolutionary Generation in Ireland 1890–1923.*
75 Mantovani, *Mazurka blu: la strage del Diana,* 160–167 and Levy, 'Charisma and Social Movements: Errico Malatesta and Italian Anarchism'.
76 Rider, 'The Practice of Direct Action: the Barcelona Rent Strike of 1931', in Goodway (ed.), *For Anarchism: History, Theory and Practice,* 79–108; Smith, 'Sardana, Zarzuela or Cake Walk? Nationalism and Internationalism in the Discourse, Practice and Culture of the Early Twentieth-Century Barcelona Labour Movement', in Mar-Molinero and Smith (eds), *Nationalism and the Nation in the Iberian Peninsula. Competing and Conflicting Identities,* 171–190; Smith (ed.), *Red Barcelona: Social Protest and Labour Mobilization in the Twentieth Century;* C. Ealham, 'An Imagined Geography: Ideology, Urban Space and Protest in the Centre of Barcelona's "Chinatown", 1835–1936', 337–397; Ealham, *Class, Culture and Conflict in Barcelona 1898–1937;* Smith, *Anarchism, Revolution and Reaction. Catalan Labour and the Crisis of the Spanish State, 1898–1923.*
77 Gorostiza, March and Sauri, 'Servicing the Customers in Revolutionary Times: The Experience of the Collectivized Barcelona Water Company during the Spanish Civil War', 908–925.
78 Guillamón, *Ready for Revolution. The CNT Defense Committees in Barcelona 1933–1938.*
79 Ealham, 'The Myth of the "Maddened Crowd": Class, Culture and Space in the Revolutionary Urbanist Project in Barcelona, 1936–1937, in Ealham and Richards (eds), *The Splintering of Spain: Cultural History and the Spanish Civil War, 1936–1939;* Casanova, *Anarchism, the Republic and the Civil War in Spain: 1931–1939,* 101–157; Lloyd, *Forgotten Places: Barcelona and the Spanish Civil War.*
80 Orwell, *Homage to Catalonia,* 4–6.
81 Preston, *The Spanish Holocaust. Inquisition and Extermination in Twentieth-Century Spain,* 221–258.
82 See the thorough engagement and comparison between the post-modernism of our era and aspects of Classical Anarchism of the late nineteenth and early twentieth centuries in the enjoyable book of Nathan Jun, *Anarchism and Political Modernity.* This is one of the themes of Ruth Kinna's new treatment of Kropotkin, *Kropotkin. Reviewing the Classical Anarchist Tradition.*

2 Uncovering and understanding hidden bonds

Applying social field theory to the financial records of anarchist newspapers

Andrew Hoyt

This chapter explores an emerging methodology for studying the transnational social networks that constituted the historical anarchist movement. Key to this approach is the examination of financial information embedded in anarchist periodicals. Analysis of the *Cronaca Sovversiva* (*Subversive Chronicle*), one of the most infamous Italian language anarchist newspapers printed in the United States during the early twentieth century, provides an excellent case study. Published in Barre, Vermont, and later in Lynn, Massachusetts, by famed orator and propagandist Luigi Galleani, the *Cronaca* circulated widely, passing through the textile mills and granite quarries of New England, the coal-camps and hard-rock mines scattered across the United States and Canada, and to over ninety locations in Italy as well as other sites of Italian immigration in France, Germany, Austria, Portugal, Brazil, Argentina and beyond. Printed between 1903 and 1918 (when it was suppressed by the US government), the *Cronaca* was a four page weekly that regularly included a financial page listing subscriptions (*abbonamenti*) and donations (*sottoscrizione*) to the paper as well as for various causes, both domestic and international, which were supported by the anarchists. By reading these lists of subscribers and donors through the lens of social field theory, scholars can gain insight into the people and communities that composed the anarchist movement. This methodology demonstrates how anarchists used newspapers such as the *Cronaca Sovversiva* as social media platforms during a turbulent era of migration and social conflict, illustrating how geographically scattered immigrant workers were able to form a powerful social movement in the face of considerable opposition.

This approach also highlights the importance of connecting local histories to larger transnational scales of analysis, thereby contextualising and connecting many seemingly isolated protests, strikes and moments of rebellion that occurred across North America in the years prior to the First World War. These previously inexplicable events, which have been explored in isolation by local historians, can now be seen as the result of actions taken by previously overlooked historical agents whose participation in a broader anarchist movement was, until now, difficult to perceive. While Galleani and some of his fellow anarchists rejected the institutional organising offered by labour unions it would be a mistake to characterise them as disorganised. Rather, for these diasporic Italian anarchists, the *Cronaca Sovversiva* and newspapers like it were key organisational tools. These newspapers facilitated the collection and distribution of resources and knit together their subscribers into what Benedict Anderson has famously called an imagined community.[1] This chapter is meant as a general call for a closer reading of the network information embedded in the hundreds of anarchist publications from this era.

Inexplicable insurrections in coal country: addressing the near-sightedness of local history

On 6 June 1909, *The Charleston Daily Mail* of West Virginia reported that Italian coal miners were refusing to follow the return-to-work orders of Ben Davis, president of the District 17 branch of the United Mine Workers of America (UMWA). Not only did these intransigent Italian miners refuse to return to work but 'as if by magic, they produced an amazing supply of rifles that they had apparently been accumulating for some time,' and proceeded to prevent over 400 fellow coal workers from returning to the mines. While the show of force used by the Italians in the area of Boomer, West Virginia (WV), was referred to as a 'riot' by local Fayette County newspapers, historian of West Virginia miners, Frederick A. Barkey, has noted that while 'in some ways these events do seem similar to spontaneous peasant revolts that were common in Italy and other parts of Europe…' nevertheless, 'there did appear to be some well thought out strategies involved'. These included gaining control of vital rail lines leading into the area and closing down central blacksmith shops that were vital to the ongoing operations of all mines in the region.[2]

By 8 June the conflict had escalated and the *Fayette Journal* reported that the miners marched on the mine offices in Boomer, 'parading behind a large red and black flag upon which was emblazoned in gold lettering the words "Victory or Death"'.[3] The fifty armed deputies who rushed to Boomer found the Italian miners 'entrenched in rock forts which they had thrown up above their homes near the Number Three mine tipple'. The crisis would remain at a standstill until another fifty armed deputies arrived the following day with a Gatlin Gun, tipping the balance of power and leading to the surrender of the insurrectionary miners.[4]

This kind of militant action was not uncommon in the West Virginia coal-fields. Indeed, in 1912, during the Paint Creek–Cabin Creek strike, Governor William Glasscock commented on the growth of radical and subversive ideologies circulating about the mine camps of West Virginia:

> The wildest theories concerning the rights of property were propounded and admitted by the strike organisers. Doctrines ranging upon anarchy were upheld with such effect that men who were before living peacefully and in comparative prosperity, purchased Winchesters, revolvers, black-jacks, and other murderous weapons to shoot down coal Barons.[5]

Governor Glasscock's description of the violent politics of the West Virginia miners was not a complete exaggeration. In fact, Governor Glasscock declared martial law during the Paint Creek–Cabin Creek strike and in July of 1912 stationed National Guard troops along Paint Creek in response to the escalating conflict between the West Virginia miners and the coal operators' thugs, particularly the Baldwin-Felts agents. Barkey states that:

> Shortly after the troops were deployed, the governor was forced to dispatch part of them to nearby Boone County where strike trouble had escalated into violence, including the wounding of Sherriff A. H. Sutphin. No sooner had troops arrived in Boone County than they were attacked at the community of Sterling by contingents of Italians from Boomer and some Greeks from the Clear Creek area... By the second week of September, the Italians walked off their jobs raising the fear that an additional 1,400 armed strikers would have to be dealt with.[6]

Not only were Italian miners known as the shock-troop of this conflict but they were also considered key to the 'logistics of keeping munitions and other vital supplies flowing into the strike zone'.[7] Once again, evidence suggests that this strike was no small undertaking. At one point, the Governor sent nine companies of troops through the strike zone and they managed to confiscate over 1,500 rifles, as well as numerous stashes of pistols, and even six or seven machine guns along with over two hundred thousand rounds of ammunition. The military authorities, apparently realising that many of these supplies had come from Boomer, sent Colonel Ford and several aides to talk to Mr Huddie, the mine superintendent, about the key leaders among the miners. Mr Huddie told Colonel Ford that the leader of the insurrectionary miners was 'an effective fellow' known as Giacomo.[8]

Historians such as Barkey have done exemplary jobs at recording this narrative of conflict, however they have often been at a loss to explain local events because they cannot see their connections to the world beyond a single town or region. In reality, events in West Virginia were not isolated but actually intimately connected to larger patterns playing out across the country and indeed across the Atlantic basin. Without a larger context, Barkey is left to conclude that, 'while it is difficult at this point to make links between specific Italian syndicalist and the left-wing West Virginia Socialists... the connection appears considerable', concluding that these Italians seemed 'infused with the anarcho-syndicalism of their homeland'.[9]

The inability to do more than make vague rhetorical gestures towards 'spontaneous peasant revolts' or the 'infusion' of politics from the 'old-country' appears throughout many local, state and regional labour histories of wild-cat militant strikes and labour conflicts. These studies seriously underestimate the anarchists' ability to mobilise working-class communities throughout the US. For example, William B. Klaus's excellent study of Americanisation amongst Italian immigrants in Marion County, West Virginia, provides great detail on similar events, but does not address the miners' motives, inspirations or goals. Klaus describes how on 15 February 1915, Italian strikers marched behind a red flag armed with weapons and carrying a banner that read 'United We Stand, Divided We Fall, Give Us Justice or Nothing at All'.[10] He goes on to comment that these miners, who lacked the backing of the UMWA or any other 'labour group' further alienated the power elite with their 'display of red flags, which was perceived as a symbol of anarchy'. Yet he concludes that:

> Only scant evidence suggests that the strikers had tangible connections to such national anarcho-syndicalist organisations as the Industrial Workers of the World or 'Wobblies'. Individual Italian radicals who organised small groups were not uncommon in coal mining regions. Perhaps such an individual was responsible for sparking the strike, but the strikers' lack of organisation suggests an unfolding of events more similar to a peasant revolt.[11]

However, if we compare his description of events to the strike described by Barkey, parallels are obvious. The protesting miners faced-off against a small posse of thirteen men who they attacked. This violent interaction resulted in the death of Constable Riggs. After this incident a second larger posse of 500 armed volunteers organised by the local authorities 'scoured the countryside, confiscating numerous pistols, knives, clubs, and one red flag'.[12] Here again we see a violent insurrection on the part of Italian coal miners that aligns with several key tropes of anarchism but is being dismissed by scholars because it lacks any clear connection to an 'official' or 'organised' union structure.

   In both of these cases of wildcat insurrectionary strikes, the ideology of the miners is explained through vague references to 'peasant revolt' and old world practices. By contrast, the methodology I use in analysing the *Cronaca* clearly shows these were not isolated events and that there was in fact a kind of organisation underlying them. These types of informal social network are commonly associated with twenty-first century social movements. However, historians have largely failed to contextualise and weave them into our understanding of historical phenomenon. By moving away from an explanatory dependency on connection to large static institutions such as labour unions to the informal structures of social networks, we can come to a new understanding of just what was happening in coal country during the years prior to the First World War. My research suggests that these miners were directly connected to one of the largest and most effective anarchist networks in the world, which was unified, educated, motivated, inspired, funded and 'organised' through the pages of the Italian-language anarchist newspaper *Cronaca Sovversiva*.

   Gossamer threads of communication and exchange ran from central nodes in the anarchist network, through the pages of circulating propaganda organs such as the *Cronaca Sovversiva*, to apparently isolated and peripheral locations such as Boomer, WV. Thus, we must start in the pages of anarchist newspapers in order to reveal the network that linked small towns like

Boomer to the larger Italian diaspora of insurrectionary labour militants. For example, the *Cronaca's* financial records show that the paper received twenty-five donations from the town of Boomer during these years. The money came from fourteen different donors and was gathered together at three separate festivals, including multiple donations from a figure named E. Di Giacomo. This fellow is very likely the same man named by Mr Huddie as the leader of the Boomer militants.

E. Di Giacomo's connection to the *Cronaca* network is underscored by the fact that he not only personally subscribed to the paper but also acted as a bundler of funds from the Boomer area. By focusing on financial records printed in the *Cronaca Sovversiva*, previously invisible base militants like E. Di Giacomo become visible as important historical actors in Boomer. We also can learn the names of some of the Italian coalminers with whom he worked. The *Cronaca's* financial section for 29 October 1910 describes E. Di Giacomo gathering together ('a mezzo') money from four other comrades to help fund the printing of a special edition of the *Cronaca* focused on commemorating the 1909 execution of the Spanish anarchist educator Francesco Ferrer. These names are all investigative leads ripe for further exploration, opening up the possibility of constructing a social history of immigrant anarchists and a more complete picture of the role they played in shaping labour relations and working-class politics during the early twentieth century.

Social field theory: thinking critically about the financial records of the *Cronaca Sovversiva*

Key to my analysis is the concept of the social field, first explored by sociologists Pierre Bourdieu and later by migration scholars such as Nina Glick Schiller.[13] Networks and social fields are closely related concepts. Glick Schiller and Peggy Levitt define social fields as

> a set of multiple interlocking networks of social relationships through which ideas, practices, and resources are unequally exchanged, organised, and transformed... Social fields are multidimensional, encompassing structured interactions of differing forms, depth, and breadth that are differentiated in social theory by the terms organisation, institution, and social movement.[14]

They argue in their 2004 article 'Conceptualising Simultaneity: A Transnational Social Field Perspective on Society' that:

> Migrants are embedded in networks stretching across multiple states... [M]igrants' identities and cultural production reflect their multiple locations. Among the important findings of the Transnational Communities project was the need to distinguish between patterns of connection on the ground and the conditions that produce ideologies of connection and

community... the nation-state container view of society does not capture, adequately or automatically, the complex interconnectedness of contemporary reality. To do so requires adopting a transnational social field approach to the study of social life that distinguishes between the existence of transnational social networks and the consciousness of being embedded in them. Such a distinction is also critical to understanding the experience of living simultaneously within and beyond the boundaries of a nation-state and to developing methodologies for empirically studying such experiences.[15]

Inspired by this analysis of the complexity of the lives and cultural production of transnational migrants, my methodology is driven by a desire to make visible and legible both the experience of migrants on the ground in small towns such as Barre, Vermont, and Boomer, West Virginia, and the production of ideologies (namely anarchism) that helped them actively build their transnational social field and diasporic community identity.

There are many different ways that migrants participate in networks and social fields. This is certainly the case with the Italian anarchists who, as we will see, were always engaged with multiple different social fields. For example, Italian-speaking immigrants used the term '*sovversivi*' (subversives) to refer to their larger community, which included Italian anarchists, communists, socialists and syndicalists alike. Thus we may describe them as part of the transnational *sovversivi* social field as well as a transnational anarchist social field that included anarchists from many different cultural and linguistic backgrounds. The readers of the *Cronaca* would also have participated in the immigrant social field (which does not imply any political focus but rather is built from shared experiences of dislocation and transnational migration among immigrants of various national origins), the United States social field (which is bound by national borders of the US) and even the Italian social field (which had a particular way of conceiving of emigrant communities as colonies and thus within the confines and gaze of the Italian state despite being legally beyond its borders). This incomplete list is meant to illustrate the numerous fluid, overlapping, intersecting and occasionally contentious identities an immigrant had to navigate on a daily basis.

One of the reasons the *Cronaca Sovversiva* makes such an interesting subject of study is that the *Cronaca* editorial group was highly conscious of maintaining financial transparency. The journal was printed weekly for over fifteen years. In total there are 779 extant editions of the paper. From this corpus I have harvested approximately 70,000 lines of data representing individual monetary transactions. The two major categories these exchanges fall under are *Abbonamenti* (subscriptions) and *Sottoscrizione* (literally 'underwritings' but best understood as 'donations'). Over the life of the *Cronaca Sovversiva*, 22,304 *Abbonamenti* and 25,908 *Sottoscrizione* transactions are listed, together comprising 48,212 of the 69,783 distinct financial transactions recorded in the *Amministrazione* section of the journal. The remaining 21,571 transactions are listed under one of the other 194 special events or causes for which the *Cronaca* raised money. Much can be done with such simple yet also extensive primary source material when read through the lens of social field theory.

First, it is important to note that not everyone who subscribed to the *Cronaca* necessarily identified as an anarchist. It is completely plausible that someone might subscribe out of curiosity or even as a way to monitor what was being said by a group that they actually deeply disagreed with. Indeed, years later (July 1918) when questioned by federal agents regarding their subscriptions to the *Cronaca Sovversiva* almost everyone named in Barre as a subscriber denied that they were anarchists. For example, Frank Juras told the federal investigators that, 'I am a subscriber to the "*Cronaca*" but am not an anarchist, and I do not believe in the principles taught by said paper. When I was a young man I did believe in some anarchistic principles but have changed my mind now' (emphasis in the original).[16] This same basic refrain is repeated over and over again to the investigators. Of course these disavowals of anarchist inclinations must be read askance, as these immigrants had every reason to lie to federal investigators, as the case of Virginio Lovargo demonstrates. Lovargo was the only one of the 25 people interviewed in Barre by the federal inspectors who admitted to identifying as an anarchist. Lovargo, age 46, willfully, if perhaps foolishly, told the agents that:

> I have been a subscriber to the 'Cronaca' from the time that it first came out in this city. I know that it is an anarchistic newspaper, and that is the reason that I read it. I AM A PHILOSOPHICAL ANARCHIST AND BELIEVE IN THE PRINCIPLES TAUGHT BY THE '*CRONACA*'. I believe that people should be educated to govern themselves and that there should be no bosses nor workmen. There should be only one class of people.[17] (emphasis in original)

This was also the only interview conducted by the investigators in Barre that ended with the ominous note '(See application for warrant for his arrest)', which highlights the potentially serious consequences of admitting such an identity. It seems surprising that an immigrant in such a precarious political and financial situation as Lovargo would make himself a target for arrest and deportation; the only sensible explanation for such behaviour is that he was so deeply tied to his anarchist identity and membership in the *Cronaca* network, that to hide, obfuscate or reject this position would represented something of a betrayal to his core values.

This series of interviews conducted by federal investigators is a reminder that while we can assume that the vast majority of subscribers to the paper at least existed within the broader *sovversivi* (Italian left) and Italian immigrant social fields in which the network operated, it is clear that we need to look to the donors' lists to see who was actually an active, self-identifying and invested member of the *Cronaca Sovversiva* network. Although subscribers received copies of the paper in exchange for their payment, there was no exchange of goods or services for donations to the network. One would not send in money to help pay off the paper's debt for any reason other than a personal interest in supporting the paper's project. In a capitalist-oriented newspaper one expects subscriptions to cover the basic cost of printing the paper, perhaps with advertisements supplementing the costs and perhaps providing a profit margin. However, in the case of the *Cronaca Sovversiva,* profit was never a driving goal, it did not solicit advertisements, and over the life of the paper donations actually brought in significantly more money than did subscriptions. In fact, at no time did money from subscriptions exceed that coming in from donations.

Additionally, money donated to the *Cronaca* arrived in bundles, meaning that numerous individual donations were gathered and sent by a single individual acting as a kind of volunteer agent for the press. US authorities recognised these acts as significant and considered anyone playing the role of a 'bundler' as having a closer relationship with the *Cronaca* group and the ideological positions they represent. For example, authorities noted in the case of Ernesto Perrella that he had 'contributed to various collections that have been taken up for its support, and has solicited and accepted subscriptions' in addition to being a regular subscriber for five or six years.[18] Perrella would later face deportation alongside with the newspaper's editor Luigi Galleani. Evidence of 'bundling' was far more damning than simply subscribing to the paper, an action whose incriminating nature the interviewees in Barre were able to deflect with ease.

In order to understand what conclusions can be made concerning people who participated to varying degrees in the circulation of the *Cronaca Sovversiva* we need to more fully employ the concept of the social field. Levitt and Glick-Schiller give us some tools to think through these relationships when they describe the difference between 'ways of being' in social fields as opposed to 'ways of belonging':

> Ways of being refers to the actual social relations and practices that individuals engage in rather than to the identities associated with their actions. Social fields contain institutions, organisations, and experiences, within their various levels, that generate categories of identity that are ascribed to or chosen by individuals or groups. Individuals can be embedded in a social field but not identify with any label or cultural politics associated with that field. They have the potential to act or identify at a particular time because they live within the social field, but not all choose to do so.... In contrast, ways of belonging refers to practices that signal or enact an identity which demonstrates a conscious connection to a particular group. These actions are not symbolic but concrete, visible actions that mark belonging... Ways of belonging combine action and an awareness of the kind of identity that action signifies.[19]

This means that, while subscribing to the *Cronaca* might simply be a 'way of being' in the social field, donating money to the *Sottoscrizione* was clearly a 'way of belonging' to the social field. Now the two kinds of acts are not exclusive or static and because of the way in which the *Cronaca Sovversiva* employed tools for identity construction such as martyrology and hagiographies, any subscriber or reader could quite easily over time come to access a sense of belonging to the social field. Commenting on this idea of belonging, Glick Schiller and Levitt state:

> Individuals within transnational social fields combine ways of being and ways of belonging differently in specific contexts. One person might have many social contacts with people in their country of origin but not identify at all as belonging to their homeland. They are engaged in transnational ways of being but not belonging...On the other hand, there are people with few or no actual social relations with people in the sending country or transnationally but who behave in such a way as to assert their identification with a particular group. Because these individuals have some sort of connection to a way of belonging, through memory, nostalgia or imagination, they can enter the social field when and if they choose to do so. In fact, we would hypothesise that someone who had access to a transnational way of belonging would be likely to act on it at some point in his or her life.[20]

Following in this same line of thinking, I argue that the list of names associated with bundling money should be read as demonstrating 'ways of belonging' to the social field and thus describe a closer degree of connection to the anarchist network. This applies to E. Di Giacomo in Boomer, WV. Conversely, we can see the *Abbonamenti* as among the lower commitment 'acts of being'. However, we should not discount the lower energy bonds of subscription or casual donation to something like the earthquake disaster fund in Calabria in 1908.

In fact, people who only appear in the pages of the *Cronaca* at these times of crisis represent a critical component of the overall social field in which the anarchists lived and operated because they showed the broadest extent of the anarchists' connections to immigrant workers and thus a kind of latent or unrealised potential energy source during moments of crisis. To see how broader and more general social fields could be mobilised by a network such as that formed through the circulation of the *Cronaca Sovversiva* we need to look at a few more case studies, starting with one which was less overtly political in nature and thus more fully capable of mobilising a broad support base among the working-class social field in towns like Barre.

The Messina earthquake case-study: being and belonging in the anarchist social field

The 1908 earthquake in Calabria and Sicily provides us with a perfect chance to explore how the anarchists mobilised the more peripheral members of the various social fields in which they belonged. Often referred to as the Messina earthquake, the 7.1 magnitude event almost completely destroyed the cities of Messina and Reggio Calabria and killed upwards of 200,000 people while sending countless more into diaspora either in Italy or abroad. The first sign of reaction in Barre seemed to have occurred on 2 January, when a general meeting of all those concerned gathered at the Northern Hotel to discuss fundraising ventures. The *Cronaca* then acted swiftly to help not only its loyal readers and fellow Italians but indeed the whole of the Barre community funnel their donations to help the people of southern Italy. The *Cronaca* did this by announcing a series of social gatherings given extra importance due to the sense of crisis the disaster had invoked throughout the whole of the Barre community and indeed much of the world.

The following week similar events were announced in other locations in the network, hinting at the extent to which the whole of the Italian diaspora was responding.[21] While it has long been noted that remittances and funds from Italians abroad played a role in helping the devastated communities recover, what is often not recognised is the role the anarchists and their networks played in this activity. The fundraising continued from early January all the way through early March, ending with a major benefit party held in New York City, which featured a scene of cross-national and cross-ideological solidarity on the left:

> The benefit "For Calabria and Sicily" given under the auspices of some anarchists and trade unionists of the upper town to the Star Casino (No. 115 E. St. 107th) could not have been more beautiful and pleasant. The crowd was a surprise, so that the take was good, and what is more important, it was a moral success. Meaning that on this painful occasion, and despite the vast theoretical gulf that divided this vast array of obscure workers, the spirit of solidarity and a brotherly hand was outstretched with sincere thoughts. It was not the usual self-interested charity of the greedy bourgeois... no! The event was spontaneous and disinterested, without any boast, as only *sovversivi* are capable...[22]

While this itself may be a boast, it is notable that the many factions of the Left came together in solidarity for the people of Calabria and Sicily. Almost all of the donors who gave to this fund did so only once, with only nine people making multiple donations, meaning 475 different people donated to the cause. The list of other donors reveals that the anarchists were successfully mobilising non-Italian donors as well.[23] The inclusion of women, children and non-Italian speaking donors suggests the degree to which this cause mobilised a much larger section of the Barre social field than typical *Cronaca* fundraising drives. This in turn requires that we ask why all these hundreds of workers in Barre gave their money to the *Cronaca* and not the Red Cross or some other more 'respected' or mainstream charity organisation.

The only possible explanation is that Barre's anarchists, despite their contentious position within the community, were respected as honest and trustworthy people when it came to handling financial donations. This trust could only have been built up over time due to their constant, successful, and fairly scandal-free fundraising activity. At moments like this, the financial transparency valued by the *Cronaca* editors proved critical. The anarchists associated with the *Cronaca* were able to handle these donations because they had access to the anarchist network which would facilitate the movement of the money and insure its delivery to the neediest with the least amount of overhead cost or loss of resources to intermediaries. Awareness for how they were able to mobilise otherwise unaffiliated members of the community at times such as this illustrates the validity of social field theory in relationship to the financial activity we find recorded in the paper. It also reveals that the fairly small network of militant anarchists involved in producing and circulating the paper actually had a much larger social field of potential supports they could draw upon when needed, notably in times of crisis as documented by some of the major strikes of Italian workers during the same years.

The 1910 Ybor strike: a study of the network's ability to respond to political crisis

Alongside *Abbonamenti* and *Sottoscrizione*, the *Cronaca Sovversiva* regularly published announcements and debriefings of small local festivals, concerts, picnics and raffles occurring in its far-flung circulation network. While these advertisements included events in major metropolitan areas such as New York City, Boston and Chicago, mostly they described gatherings, dinner parties, theatrical recitals and other 'happenings' that animated life in Italian communities in coal camps and small villages of the American west. Through a close examination of the paper we can observe how the announcement of one locality's decision to stage a fundraising and community building activity for a particular cause would spread in both form and function to other communities. This kind of snowballing of activity was not accidental or peripheral to the anarchist movement or the role that newspapers played in it, but represented one of the key ways anarchist periodicals facilitated mobilisations during the first two decades of the twentieth century.

For example, in June 1910, the Clear Havana Cigar Manufacturers Association began firing Union Selectors from the Cigar Manufactures International Union (CMIU). By August more than 12,000 cigar makers were out of work.[24] Alfonso Coniglio, one of the strike's primary leaders, was a major anarchist leader and distributor of the *Cronaca Sovversiva*.[25] Indeed, records show him contributing to the paper from 1903–1917, as the second most active subscriber and donator from the Florida area, associated with the Alba Sociale Gruppo in Ybor City. By September, conflict was increasing and an Anglo-American bookkeeper was shot in the midst of a fight with a crowd of strikers. Two Italian men, Angelo Albano and Castronse Figarretta, were labelled 'tools of anarchist elements in the city' and were lynched. Cigar factory works of Italian, Spanish and Cuban descent moved from protest to acts of violence.[26] On 4 October arsonists burned the Bulbía Brothers factory.[27] Cigar factories reopened October 1910 with armed protection. Italians refused to return to work, marching behind a banner similar to that reported in West Virginia that read '*Morire di Fame, ma Vincere!*' (We Will Die of Hunger, But We Will Win!). Italians once again became the shock-troops of the strike.[28]

The *Cronaca* regularly reported on and supported such strikes. It became highly active in raising funds for strikers and in publicising their struggle. The paper's reaction to events in Ybor City provide a good case study of how it mobilised far-flung and diverse communities of sympathetic workers. Between 31 December 1910, and 11 March 1911, twenty special fundraising events were reported in the pages of the *Cronaca*. Twelve of these festivals are explicitly for the strikers in Tampa. The first two 'Festa di Ballo' occurred in Mulberry Kansas, and Bay View Massachusetts, on 31 December. In total there were eighteen reports (*communicate*) from towns announcing plans for an event and reporting back on how the event unfolded, and how much money was raised. For example, the report from Coalgate, Oklahoma, stated that tickets sales brought in $71 dollars, cigar and soda sales brought in eight more but the cost of renting the hall and musicians was $52.38, bringing their total profit to $27.50.

There are also eleven articles about the strike published in the paper at this time, including six large front page articles starting on 12 November and running until 31 December, when news from Tampa fell to the second or third page of the paper. These articles offer detailed analysis of the conflict, as well as rhetorical calls for solidarity and support. When combined with the stream of reports on festivals, such press would certainly have made any reader not only aware of the lynching, beatings and oppression faced by the Tampa workers but also the way in which other Italians, in the coal camps of the mid and far west, the granite quarries of the north, the urban streets of New York and New Jersey and the factory towns of New England were responding to the brutality meted out to their fellow workers. Such a complicated set of representations and communications would not only have moved a reader to outrage but given them specific ways of venting their outrage and participating in the struggle.

Cronaca printed accounts of total financial contributions from sixteen festivals, picnics, raffles and group and individual *sottoscrizini* in support of the Tampa strikers between 26 November and 11 March. This running total also included information on how much money has at any time been sent to Tampa. It was often accompanied by printed copies of receipts showing that funds were received by the strike committee in Florida. In this way the *Cronaca* built a reputation for transparency that allowed the paper to function as financial intermediary for a broader movement.

On the one hand this openness could be seen as surprising (given the high-level of secrecy common among insurrectionary anarchists in other arenas), but also understandable given the fact that workers were giving significant portions of their insubstantial, if hard-earned wages. These financial records are often the only historical records left by these men and women, migrants who may only otherwise appear in INS files, census records and police documents. The pages of the *Cronaca Sovversiva* offered a space for individuals to signal to themselves and to their fellow readers their membership in the movement. Also having one's name printed in the paper, not only as a subscriber but as someone sending in extra money to support a specific cause, such as the strikers in Tampa, would have certainly been a source of pride for an autodidact and subaltern coal miner scattered across the American West. A fact that would change as the anarchist social field came under attack at the end of the decade and all those subscribers in Barre were forced to repudiate association with the *Cronaca* network.

Money for the strikers continued to arrive in *Cronaca* offices until the end of the Tampa strike, when workers agreed to return to factories on 26 January 1911. At this time a momentary drop off of incoming funds was followed almost immediately by a very high spike, as all the groups who had been raising funds sent in whatever they had gathered to support the strikers. The surge in donations after the collapse of the Tampa/Ybor strike suggests that the anarchist network knew strikers would especially need assistance after the failure of their struggle. Such generosity from their fellow workers must have provided some emotional consolation, as well as some food and rent money for the defeated strikers and their families. This act undoubtedly strengthened the bonds between the diasporic groups of workers and increased loyalty to the *Cronaca* and its anarchist platform. The workers of the Tampa/Ybor area did not forget the role anarchists and newspapers like the *Cronaca* played in supporting them during their strike, and over the following years funds streamed out of the cigar factories and into anarchist journals.[29] In fact, we can see the money coming from Tampa into the *Cronaca* spike in 1911 and slowly decline for several years afterwards, as the memory of the strike and the support given to the strikers slowly faded from the collective memory.

Conclusion

By paying close attention to the flow of money through the pages of the anarchist press we can learn more about the ways in which migrant labourers fought back against exploitation. An examination of this financial data reveals a pattern of fundraising that bound together working-class radicals across North America and the Atlantic basin. The anarchist social network must be understood as a form of organising that was coexistent with other articulations of working-class self-empowerment, such as the union movement, but that was radically different in its approach to grassroots mobilisation and resistance. The financial records of papers like the *Cronaca Sovversiva* reveal that the organisational strength of immigrant Italian anarchists lay not in their numbers or in the formation of lasting institutions such as unions or political parties but in their highly flexible networks which allowed small numbers of militants to mobilise sympathetic (though not necessarily radicalised) members of the larger immigrant and working-class social fields. Viewed from this perspective, the insurrections led by coal miners in West Virginia were not isolated events or 'spontaneous peasant revolts', but were part of a larger phenomenon.

When anarchists began to organise fundraising events for the victims of the Messina Earthquake, many members of the broader community trusted them to gather and deliver aid to Italy rather than turning to alternatives such as the Red Cross. Because of their well-earned reputation as effective grassroots fundraisers and honest redistributors of wealth, they found themselves in the position to respond immediately to crisis and used the opportunity to mobilise people beyond the inner core of their social field. It is this ability to mobilise that made the anarchists so resilient to attack. The hundreds of times they had previously delivered funds, as they did for the striking cigar rollers of Tampa, had built up a wealth of social capital. Their reputation extended the reach of their network farther into the immigrant social field and helped them react to disasters both natural and political. This was not an accidental consequence of their anarchism but a direct result of their ideology of social revolution, which stressed working-class self-reliance.

These short studies address only a fraction of the information contained in the financial records of the *Cronaca Sovversiva*. And the *Cronaca's* records represent only a tiny fraction of the information contained in the print culture of the anarchist movement writ large. If more of this data is transcribed and made available to the scholarly community we will be able to build a more complete map of the flow of radical ideology, financial resources, and subversive identities during the height of the historical anarchist movement. In this way we can move beyond intellectual history and biography to build a social and cultural history of the previously nameless labourers whose struggles for social justice shaped not only the past, but also the world in which we live today.

Notes

1 For more information see Anderson, *Imagined Communities: Reflections on the Origin and Spread of Nationalism*.
2 Fones-Wolf and Lewis, *Transnational West Virginia*, 169.
3 Fones-Wolf and Lewis, *Transnational West Virginia*, 168.
4 Fones-Wolf and Lewis, *Transnational West Virginia*, 170.
5 Fones-Wolf and Lewis, *Transnational West Virginia*, 179.
6 Fones-Wolf and Lewis, *Transnational West Virginia*, 174.
7 Fones-Wolf and Lewis, *Transnational West Virginia*, 174.
8 Fones-Wolf and Lewis, *Transnational West Virginia*, 174–175.
9 Fones-Wolf and Lewis, *Transnational West Virginia*, 180.
10 Fones-Wolf and Lewis, *Transnational West Virginia*, 191.
11 Fones-Wolf and Lewis, *Transnational West Virginia*, 194–195.
12 Fones-Wolf and Lewis, *Transnational West Virginia*, 195.
13 For more information see Bourdieu, 'The Social Space and the Genesis of Groups'.
14 Levitt and Schiller, 'Conceptualizing Simultaneity', 1009.

15 Levitt and Schiller, 'Conceptualizing Simultaneity', 1006–1009.
16 John W. Dolan, 'Investigation in Re Anarchists at Barre, VT', Investigation (Barre, Vermont: Immigration and Naturalization Service, 31 July 1918), 5, Box 2801; Subject and Policy Files, 1893–1957; Records of the Immigration and Naturalization Service, Record Group 85, National Archives, Washington D.C.
17 Dolan, 'Investigation', 7. Lovargo also commented that he was married but had never had a marriage ceremony because 'I am not defying the law of any state, but I do believe that marriage is not necessary when you can get a divorce. We married ourselves; there was no clergyman or justice of the peace needed.'
18 John M. Lyons, 'Brief Submitted on Behalf of the Commissioner of Immigration, in Re Habeas Corpus Petition of Luigi Galleani, et Al.', Brief (Boston, MA: Immigration and Naturalization Service, n.d.), Box 2801; Subject and Policy Files, 1893–1957; Records of the Immigration and Naturalization Service, Record Group 85, National Archives, Washington D.C.
19 Levitt and Schiller, 'Conceptualizing Simultaneity', 1010.
20 Levitt and Schiller, 'Conceptualizing Simultaneity', 1010–1011.
21 [Anon.] 'Per le vittime del terremoto', *Cronaca Sovversiva*, 16 January 1909, 4.
22 Gigione, 'I sovversivi e il terremoto', *Cronaca Sovversiva*, 6 March 1909, 3.
23 The names of those who gave twice appeared fairly commonly in the *Cronaca* and thus represented important players in the anarchist network. None of the non-Italians appeared again in the *Cronaca Sovversiva*, and those that did, such as Albert Halvosa, do so because they were prominent leaders of the stone carvers' union and usually were seen as antagonists by Galleani and the other anarchists.
24 Mormino and Pozzetta, *The Immigrant World of Ybor City Italians and Their Latin Neighbors in Tampa, 1885–1985*, 119.
25 Mormino and Pozzetta, 'The Radical World of Ybor City, Florida', in Cannistraro and Meyer (eds), *The Lost World of Italian American Radicalism: Politics, Labor, and Culture*, 256.
26 Mormino and Pozzetta, *The Immigrant World*, 120.
27 Mormino and Pozzetta, *The Immigrant World*, 120.
28 Mormino and Pozzetta, *The Immigrant World*, 121.
29 Mormino and Pozzetta, 'The Radical World', 257.

3 The other nation
The places of the Italian anarchist press in the USA

Davide Turcato

Introduction: What does 'Italian' mean?

The concept of 'Italian anarchist press in the United States' is deceptive in its seeming simplicity. To start with, what does 'Italian press' mean? References to nations, in the absence of further qualifications, are usually understood as references to nation-states. So, for example, if we simply talk about 'English press' or 'Spanish press', we are surely referring to the press of England or Spain. However, if we talked about 'English press in the United States' or 'Spanish press in the United States', we would be open to two interpretations with quite different meanings: Would we be referring to the nation or to the language? Would we be talking about people who look upon England or Spain as their own country or simply about English or Spanish speakers? One could argue that, in the Italian case, we would be talking more or less about the same group of people, but the fact remains that we would be talking about it from two different points of view.

The different implications of either interpretation become clear if we consider the equally deceptive concept of 'Italian anarchist press'. While the 'linguistic' interpretation, as 'anarchist press in Italian', would be quite natural, the 'political' interpretation would place anarchists in a tight spot, for acknowledging and asserting their national identity would immediately seem to put them in contradiction with their alleged cosmopolitanism, or would make it seem a hollow abstraction, anyway.

Yet it is precisely to the latter interpretation that I will adhere. I intend to survey the geographical distribution of the Italian anarchist press in the United States, its areas of circulation, the flow of its financial support, and the mobility of its personnel, from the anarchist movement's start to the eve of the Second World War, in order to show that its periodicals were made by men and women who looked upon Italy as their country and did so consistently with their cosmopolitanism and internationalism.

The geography of the Italian anarchist press in the United States

The geographical distribution of the Italian anarchists in the United States is itself evidence of their link with their nation. If we consider the distribution of

Table 3.1 Distribution of Italian anarchist periodicals by state, 1872–1971

State	Division	Periodicals
New York	Northeast	33
Illinois	Midwest	11
New Jersey	Northeast	8
Massachusetts	Northeast	7
Pennsylvania	Northeast	5
California	West	5
Florida	South	4
Vermont	Northeast	3
Ohio	Midwest	2
Minnesota	Midwest	1
Connecticut	Northeast	1
Rhode Island	Northeast	1
Total		81

Source: Bettini, *Bibliografia dell'anarchismo*, vol. 2

the press an indicator of the anarchist movement's distribution, we clearly see how the latter reflected the distribution of Italian immigration. Table 3.1 shows the distribution of Italian anarchist periodicals per state, up until 1971. The data are taken from Bettini's *Bibliografia dell'anarchismo*, which is the main reference source for any research on the Italian anarchist press. The Northeast region clearly predominates with its 58 periodicals (71.6 per cent), followed at a distance by the Midwest with 14 periodicals (17.3 per cent), the West with 5 periodicals (6.2 per cent), and the South with 4 periodicals (4.9 per cent). This distribution mirrors the distribution of Italian immigration, which in the last decade of the nineteenth century was accounted for 72.7 per cent by the Northeast, for 11.4 per cent by the Midwest, for 8.3 per cent by the West, and for 7.6 per cent by the South.[1]

However, Bettini's data provide only a partial image. The anarchists' distribution and that of their press did not depend exclusively on the volume of Italian immigration, but also on other and more specific dynamics. In some divisions, such as the South Central – especially Louisiana – the anarchist press was absent, although Italian immigration was higher (5.4 per cent in the abovementioned decade) than in other divisions, such as the South Atlantic, that had a lower Italian immigration but provided a more fertile ground for transplanting traditions of radicalism imported from Italy.

We begin to discern a different picture if we replace statistics based on the number of periodicals – where equal weight is given to a one-off publication and a decade-long periodical – with statistics that take into account the life span of a periodical, measured in weeks, with a minimum span of one week

Table 3.2 Cumulative life span of Italian anarchist periodicals, grouped by state

State	Weeks
New York	5,508
New Jersey	1,547
Massachusetts	1,314
California	682
Vermont	572
Pennsylvania	411
Florida	140
Illinois	113
Ohio	73
Minnesota	1
Connecticut	1
Rhode Island	1
Total	10,363

Source: Data from Bettini, *Bibliografia dell'anarchismo*

conventionally assigned to one-off publications (Table 3.2). We can thus reassess the relevance of the densely populated Illinois, with its numerous but ephemeral periodicals, while faraway California and the small and mountainous Vermont gain importance, with their few but long-lived papers.

An even sharper picture takes shape if we introduce a temporal distinction, with the Great War as watershed (Table 3.3). Two surprisingly different scenarios emerge. The metropolitan state, New York, which seemed to dwarf all other states with its 41 per cent of periodicals over the entire period, took on this role, and impressively so, only after the war, accounting for 70 per cent of the entire production. However, before the war its role is marginal (10 per cent), whereas we have a confirmation of a marked decentralising tendency toward minor states, such as New Jersey and Vermont, which together account for 61 per cent of the production.

We can further detail the pre-war situation by aggregating the same data by city rather than by state (Table 3.4). The interesting fact is that eight cities alone suffice to cover 99 per cent of the entire production. The tendency to decentralisation in terms of states thus corresponds to a tendency to concentration in terms of cities. A single place corresponds to each of the previously listed states: New Jersey actually means the town of Paterson; Vermont means the town of Barre.

At the root of this phenomenon there is probably still migration, in one of its most characteristic aspects, chain migration, the result of which was that immigrants tended to cluster according to their place of origin. This phenomenon favoured migration between areas where the same kind of economic

Table 3.3 Periodicals' cumulative life span, by state, before and after 1915

Before 1915		After 1915	
State	*Weeks*	*State*	*Weeks*
New Jersey	1,173	New York	5,217
Vermont	572	Massachusetts	1,099
Pennsylvania	300	California	492
New York	292	New Jersey	375
Massachusetts	214	Pennsylvania	111
California	190	Florida	98
Illinois	81	Ohio	72
Florida	42	Illinois	31
Ohio	1	Connecticut	1
Connecticut	0	Rhode Island	1
Rhode Island	0	Minnesota	1
Minnesota	0	Vermont	0
Total	2,865	**Total**	7,498

Source: Data from Bettini, *Bibliografia dell'anarchismo*

Table 3.4 Periodicals' cumulative life span, by city, before 1915

City	Weeks	Cumulative coverage	
Paterson, NJ	1,173	1,173	41%
Barre, VT	572	1,744	61%
Philadelphia, PA	300	2,044	71%
New York	292	2,336	82%
Lynn, MA	213	2,549	89%
San Francisco, CA	190	2,739	96%
Madison, IL	43	2,783	97%
Tampa, FL	42	2,825	99%
Other cities	40	2,865	100%

Source: Data from Bettini, *Bibliografia dell'anarchismo*

activity was prevalent; this, in turn, had the consequence that worker radicalism typical of certain Italian areas of origin concentrated in corresponding areas of destination, where overall immigration may have been lower than elsewhere. Thus Paterson, with its silk mills, attracted workers from the Biellese, where the textile industry tradition harked back to the Middle Ages and where great strikes took place between 1877 and 1889. Analogously, the Barre anarchists were quarrymen coming from the area of Carrara, the city of the 1894 anarchist uprising. Another example is Tampa, Florida, where Italian immigration was mostly from Sicily, the scene of the 1893 Fasci movement, and where the tobacco industry was characterised by strong worker combativeness.

The concentration in terms of cities – with different periodicals following one another in the same towns, many of which were not even densely populated – invites us to think about the type of infrastructures that made an anarchist periodical possible. The two key requisites were an active anarchist group and a printing press, both of which could equally be found in a large city and in a small town. An anarchist periodical was never an end in itself, but was always instrumental to a larger project of dissemination of political ideas. The places where anarchists set up their periodicals were, or would become in time, organisational hubs to which various activities were connected. On the one hand, the most representative periodicals turned into actual organs of the movement, which took on the function of correspondence committees that kept up contacts within a network of comrades spread over a large territory. Their standard set-up included not only an editor but also a public speaker. On the other hand, printed materials for the dissemination of ideas was not limited to periodicals, but almost invariably included also pamphlets, posters, etc. The printing presses that produced such literature were often more durable than the periodicals themselves, and their presence in a certain town was often a key factor that made it possible for a new periodical to be established or to flourish. A significant example is that of a Spanish anarchist and professional typesetter, Pedro Esteve, who, in 1899, moved his typing equipment from New York, where he published a periodical in his own language, to Paterson, so that it could be used by three other anarchist periodicals – two Italian and one French – in addition to his own. A steady group and a printing press were thus the two factors that provided continuity to anarchist literature.

In the light of the above survey, we can now give a name to the main periodicals, mentioning the four longest-lasting ones: *La Questione Sociale* (The social question) of Paterson and its reincarnation, *L'Era Nuova* (The new age); *Cronaca Sovversiva* (Subversive chronicle) of Barre; *Il Martello* (The hammer) of New York; and *L'Adunata dei Refrattari* (The refractories' call), also based in New York. The foregoing overview lets us better appreciate how important and representative these periodicals were. Their places of publication are significant in themselves: the first two periodicals are the chief representatives of the pre-war period, while the last two represent the post-war period. These periodicals also epitomise another distinction, in terms of ideas, that ran through the anarchist movement: the distinction between 'organisationists' and 'anti-organisationists'. The debate between the two camps did non concern only formal organisation in federations that drafted programmes and held congresses, as was favoured by the former side and opposed by the latter, but also other issues such as the so-called 'propaganda by the deed', participation in the labour movement, and alliances with other subversive parties. Among the abovementioned periodicals, *La Questione Sociale—Era Nuova* and *Il Martello* represented the organisationist camp, although with different nuances, while *Cronaca Sovversiva* and *L'Adunata dei Refrattari* represented the anti-organisationist camp. These four periodicals alone cover 59 per cent of the cumulative life span of all periodicals (6,077 weeks out of 10,363). Therefore they constitute the backbone of the Italian anarchist press in the United States. Accordingly, we will make reference mainly to them in the rest of our discussion.

The role of transnationalism

The prize of first Italian anarchist periodical in the United States, however, is carried off by a short-lived paper, *L'Anarchico* (The anarchist), which saw the light of day in New York in 1888. The front page of its second issue (1 February 1888) carried an appeal 'Ai Compagni d'Italia' (To the comrades of Italy), which explained that the periodical's publication was undertaken 'in view of the fact that it is presently impossible to express one's own ideas, because of the zeal of the tyrannical Italian police, in the pay of a constitutional monarchical government'. Moreover, the editing group appointed one of its members 'so that he can be recognised by all the Socialist Anarchist Revolutionary groups to which he will introduce himself in Italy, for the purposes of giving and receiving whatever explanations he will be asked for, facilitating subscriptions, and obtaining direct reports from the groups of Italy'.

The appeal is highly significant. In symbolically marking the birth of the Italian anarchist press in the United States, it also constitutes, as it were, its manifesto. The Italian anarchists of the United States considered themselves part and parcel of the movement in their homeland: their press was primarily made to serve that movement. That press bore the advantage that it could be produced in easier conditions, and therefore its role became especially important in times of repression in Italy. Such was the division of tasks between Italian anarchists at home and abroad.

We can find such concepts explicitly expressed in the anarchist papers. For example, in a September 1899 issue of *La Questione Sociale*, Errico Malatesta – the most representative figure of Italian anarchism and the editor of that paper at the time – emphasises that economic and political conditions are more favourable in the United States than in Europe, and therefore 'we must profit from the circumstances to create a force that can ... come to the aid of our cause wherever the opportunity arises, and especially in Italy, which is the country we come from, whose language we speak, and where, consequently, we can exert our influence with greater effectiveness'.[2] We can see here how the priority given to one's own country did not mean preferring it to other countries, but rather proceeded from a sort of internationalist division of labour, based on the fact that each one's action could be more effective in the country whose culture one shared.

An indication of the greater difficulty of survival of the anarchist press in Italy than in the United States comes from the comparison of the number of one-off publications in each country. If we consider a single issue as the ephemeral publication par excellence, the percentage of single issues over the total number of publications indicates the precariousness of a country's press. In Italy such ratio was 59 per cent (175 periodicals over 298) in the period up to 1913, in contrast to a ratio of 41 per cent abroad (54 periodicals over 131). In the United States the ratio was still lower than the overall ratio for foreign countries: 34 per cent (10 periodicals over 29). In this last country the percentage remains approximately constant throughout the entire period up to 1971.

A further indication of the different conditions in the two countries – as well as an indication of the role of the anarchist press in the United States in times of repression – comes from the respective year-by-year fluctuations in the number of periodicals between 1893, the year in which the Italian anarchist press started to have a steady presence in the United States, and 1927, the year in which the anarchist press disappeared from Italy, as a result of government suppression of all antifascist press (Figure 3.1). The number of periodicals for each year was conventionally observed on the 1 January of the year. The average number of periodical per year in Italy is exactly twice the US average (5.66 and 2.63, respectively).

Figure 3.1 lends itself to various considerations. First of all, the press in the United States had a more regular course than the press in Italy, which was more exposed to government repression. This can be observed by comparing each country's series of data with the straight line that best fits the data (also shown in the figure). For example, the line shown for Italy in Figure 3.1 has a value of 4.7 for 1894, which, compared with the value of 6 periodicals published in Italy that year, yields a deviation of 1.3. The best-fit line, also called 'trend line', is the one that minimises the average of such deviations (expressed as

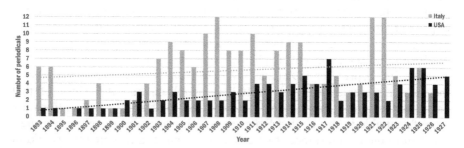

Figure 3.1 Yearly fluctuation in the number of periodicals in Italy and United States, 1893–1927

Data from Bettini, *Bibliografia dell'anarchismo*

absolute numbers, irrespective of their sign). The closer the series of data is overall to the line of best fit – that is, the lower the average deviation from that line is – the more regular the course is. The average deviations for Italy and the United States are 2.9 and 0.8, respectively. However, a meaningful comparison should factor out that the two data series have different scales, as we usually expect higher deviations for series made up of larger numbers. We normalise the average deviations by dividing them by the respective data series averages. The resulting average deviation for Italy is 51.4 per cent of its average number of periodicals per year, while the corresponding value for the United States is 28.5 per cent. In brief, as the chart visually shows, the fluctuation for Italy was much more jagged than that for the United States.

Even more notably, the respective fluctuation of the Italian anarchist press in Italy and the United States often went in opposite directions. At various times, when the volume of publication was below the trend line in Italy, it exceeded the trend line in the United States. We observe this for the years 1900–1901, 1916–1917, and 1924–1925. As a result, in these periods the number of Italian periodicals in the United States equalled or even exceeded the number of periodicals in the homeland. We also observe the inverse phenomenon: when the anarchist press flourished in Italy it dropped below the trend line in the United States, as we can see for the years 1905–1908 and 1921–1922.

These statistical patterns acquire significance in the light of the historical events Italy went through over those nearly four decades. After Crispi's exceptional laws are promulgated in 1894, the anarchist press drops in Italy. In 1896 we thus see for the first time the United States press on a par with the Italian press, albeit at the lowest level. Then, in the years of the repressive backlash ensuing the 1898 bread riots and Gaetano Bresci's 1900 attempt on King Humbert's life, the United States press steps in and for the first time exceeds homeland production. The central years of the Giolittian era, during which the grip on the subversive press was slackened, see the anarchist press soar in Italy and at the same time mostly fall under the trend line in the United States. With the outbreak of the First World War, the anarchist press has a hard time in Italy but it grows in the United States, where in 1917 there are nearly twice as many periodicals as in Italy. In the wake of the 1919–1920 Red Biennium, production grows to a record high in Italy, while across the Ocean it hits a record deviation below the trend line. However, after the rise to power of the fascist regime, the anarchist press quickly drops in Italy, until it is completely silenced in 1927, while in the United States it rises with equal rapidity above its trend line, to remain one of the main voices of Italian anarchism for several years to come. In sum, a clear picture emerges, which shows that the Italian anarchist press in the United States consciously took on the fundamental role of giving voice to Italian anarchism whenever that voice was choked in the homeland. When that urge was less pressing, Italian anarchists in the United States were less motivated to step up their production and were rather inclined to channel their resources towards the press in Italy.

A few paradigmatic episodes may help further highlighting that role. The precarious condition of the anarchist press in Italy during the First World War is vividly illustrated by the front page of the 2 March 1916 issue of *Il Libertario*, a paper from the Ligurian city of La Spezia (Figure 3.2).

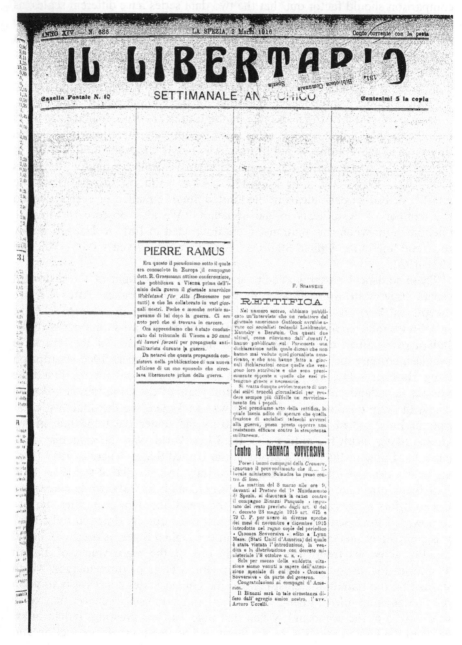

Figure 3.2 Front page of *Il Libertario* (La Spezia), 2 March 1916

Among the sparse short articles that survived censorship in that page, there is one that reports the forthcoming trial against the *Libertario*'s editor, Pasquale Binazzi, on a charge of having repeatedly imported into the Italian kingdom copies of the periodical *Cronaca Sovversiva*, 'the importation, sale, and distribution of which were banned by ministerial order of the 12 October ultimo'. What led the government to ban that periodical and the anarchists of Italy from smuggling it into the kingdom is illustrated with equal vividness by a front page of *Cronaca Sovversiva* published a few weeks before the ban (Figure 3.3), which, under the sarcastic title of 'Evviva la guerra!' (Long live war!), gave free expression to the antimilitarist ideas that were being gagged in Italy.

In times of strongest censorship it was common practice for the most prominent personalities in the movement, such as Malatesta, to simultaneously send the same articles to the press of the United States and of Italy – a further indication of how the press of those two countries addressed the same readership. What was different was the outcome of the respective editions. For example, in May 1916 both *L'Era Nuova* of Paterson and *Il Libertario* of La Spezia printed an important antimilitarist article by Malatesta, 'Anarchici di governo' (Pro-government anarchists), both giving it the leading article's pride of place. Unsurprisingly, the former was published as fully as the latter was blanked by the censors. Clearly, such a state of affairs intensified with the rise of Fascism. In 1931–1932, Malatesta's last years of life, during which he lived in Rome in an undeclared state of house arrest, it was only through New York's *Adunata dei Refrattari* that his voice could still be heard.

During the Fascist regime the anarchists shared such a situation with all other antifascist parties. The importance of emigration and antifascist press abroad in that period is well-known and therefore it may seem superfluous to specifically emphasise it in the case of the anarchists. What is important to stress, however, is that in the case of anarchism – at different degrees in different times – this was the standard modus operandi. Fascism did not so much represent an exception, a dark interlude, as the intensification of a censorial practice to which anarchists were accustomed.

Transnational readers

It is difficult to provide systematic figures about the circulation of anarchist periodicals. Table 3.5 summarizes the available data, sorted by highest circulation.

The few periodicals that reach or exceed the 5,000 copies are all later than 1915, with a peak reached by the New York *Il Martello*, the champion of antifascism and of Sacco and Vanzetti's defence, edited by Carlo Tresca. Otherwise, a circulation of about 3,000 copies seems to be the standard for periodicals that managed to live longer than a year. More detailed data concerning *La Questione Sociale* seem to confirm this generalisation. We know that in February 1899, when the paper was published weekly, its circulation grew from 3,000 to 3,500 copies. The circulation was further increased to

ANNO XIII LYNN, MASS. 11 SETTEMBRE 1915 Num. 37

CRONACA SOVVERSIVA
EBDOMADARIO ANARCHICO DI PROPAGANDA RIVOLUZIONARIA
UT REDEAT MISERIS ABEAT FORTUNA SUPERBIS

Corrispondenze, lettere, money orders debbono essere esclusivamente indirizzati "Cronaca Sovversiva", P. O. Box 678 – Lynn, Mass.

EVVIVA LA GUERRA!

È, fra i guerrafondai, della gente seria, che vi persuade.

Non fra i Tirtei del nazionalismo strillone, badiamo bene! Non fra i Rostand, i Kypling, i D'Annunzio che sulla lira malconcia — ridotta oramai al solo monotono episodio della grande guerra fascinatrice — vi chieggono pel buon dio e pel re, per la civiltà e per la patria, discretamennte, la pelle, il sangue e la giovinezza dei figlioli, le lacrime di tutte le madri; e vi persuadono tanto meno che il fervore, l'estro e l'entusiasmo essi dànno in affitto a chi meglio paghi.

Bisogna d'altra parte riconoscere che hanno per le mani così aspra bisogna cui non basterebbero nè il genio nè il plettro nè gli anni d'Omero.

Persuadere al proletariato che per la causa della civiltà — a cui esso è rimasto fin dalle origini, traverso i millennii, straniero immutabilmente — egli non ha dato fino ad ora che una miseria: **cinque milioni duecentonovanta mila [illegible], quattordici milioni trecentonovantotto mila perduti** in totale, nel primo anno di guerra 1); e che deve darne ancora, in proporzione maggiori, per altri due o tre anni, finchè le sorti del progresso e della libertà non siano vittoriosamente decise, non è compito del resto che si possa onestamente pretendere dai giullari di corte; nè che si possa risolvere con un penna, con una fanfara, con quattro chiacchere garbate.

⚜

La gente seria non fa tanti discorsi, ha in eguale dispregio il ciarpame venerando delle tradizioni come le dubbie promesse dell'ideale, così remoto fra le nebbie dell'avvenire quanto è dubbia e lontana nella tenebra dei tempi la scaturigine degli orgogli e dei miti della stirpe. Non crede che all'oggi, alla realtà immediata, concreta, ponderabile; e la coglie e la pesa nel minuto che passa, al di là del quale essa è completamente diversa, così diversa che è orrore domani quel che oggi è gloria, rovina ed onta quello che oggi è fasto o dovizia.

Vivere l'ora! e viverla piena ed intiera, con la tensione di tutti i nervi, di tutta la volontà, di tutta la forza, al fine limpido ed immediato; e la meta attingere coll'impeto subitaneo del ciclone, travolta ogni barriera del sentimento e della ragione, della giustizia e della pietà, nel dominio, nell'onnipotenza dell'attimo fugace.

Poi? domani? dopo?

1) Sono le cifre, discrete, riassunte dal Ministero francese della guerra al 31 Maggio 1914.

Venga il diluvio; altri andrà sommerso. Essi non vi saranno più.

⚜

Non è così la gente seria?

Misuratela alla stregua della grande guerra.

Non le ha mai cercato una giustificazione all'infuori del calcolo cinico e brutale: la guerra urge! urge sbucare dalla mediocrità e dal marasma, a non asfissiare tra il pidocchiume, a restituire agli arrembaggi ed alle corse l'antica maestà soggiogata dalle convenienze, dai trattati laboriosi, accidiosi, subdoli o taccagni; a restaurare sopratutto l'epica grandezza e la dovizia tragica del sacco e del bottino: urge a ricondurre nelle ciurme, che farneticano briache di diritti e di rivolte, la fedeltà, la disciplina, la soggezione.

Ha buttato fra pretoriani citaredi e giullari della chiesa e della caserma, del parlamento e dell'Ate-[illegible] d'oro e l'ordine sovrano egualmente irresistibile: [illegible] benedissero i soli nel nome del diritto, la sancirono i parlamenti nel nome della stirpe, le coscrissero i cortigiani nel nome del re gli eserciti fra le ciurme tornate nel nome della legge e della patria dallo scisma pavido e discreto alle cieche sudditanze sotto i vecchi labari riconsacrati.

Dio e la fede, il re e la legge, il diritto e la patria hanno dato l'aureola, il viatico, le armi e gli entusiasmi di una crociata alla cinica avventura, a l'editto bestiale che in origine, spogliati degli orpelli bugiardi, dicevano semplicemente: **sgombrate il mercato dalla marmaglia taccagna che qui si hanno a coniare quattrini, e restaurare su le disperse trincee dell'avvenire le provvidenziali bastiglie dell'antico regime.**

⚜

S'è urtato il proposito ad ostacoli insospettati e, trascendendo ogni più temeraria previsione, la rovina ha levato da tutti i cuori lo sciame sgomento d'inquietudini disperate: "Nel gorgo le cattedrali e le fortune; a fiotti, nel gorgo, sudor di fronti, tesori di fatiche, d'ingegni, d'invenzioni, d'ardimenti millennarii, nel gorgo incommensurato i tesori del domani, le volontà più fervide, il sangue più generoso, le energie più agguerrite, l'inapprezzabile necessaria guarentigia della vita e del progresso: la minaccia d'un più fosco millennio sul conserto destino delle genti... Troppo, troppo!"

E la diana squillante ieri su per la gamma degli entusiasmi universi, le promesse della vittoria fulminea e decisiva, la magica rinnovazione della vita, il trionfo delle audacie giovanili, della forza turgida incoercibile, [illegible] su l'inutile desolazione, tra la [illegible], l'orrore e la paura, in singulti disperati da cui gorgogliano la protesta e la maledizione.

⚜

La gente seria [illegible]; squaderna su gli [illegible] il libro mastro, l'indice [illegible] su la colonna: "Profitti e perdite".

Rovina e desolazione? Dopo un anno di guerra prodigalmente vorace zampilla a torrenti l'oro dalle terre devastate.

Cinquecento milioni in oro si rovesciano dai galeoni britannici nei forzieri di J. P. Morgan a dirvi l'immensità riparatrice delle taglie che li riscatteranno domani, mentre dal sangue degli olocausti propiziatori s'irrorano, miracolosamente ringiovanite, fortune decrepite e moribonde.

"[illegible] cristianissimi della pietà, che vi pascete di brivi-[illegible] nostre industrie anemizzate? La più potente delle nostre organizzazioni metallurgiche la Bethlehem Steel Trust Co. non trovava nel gennaio scorso un cane che volesse delle sue azioni a quarantasei dollari ed un quarto.

Urge la guerra, soggioga tutte le braccia alle armi, tutte le officine del vecchio mondo al vassallaggio della nostra intraprendenza, e le azioni della Bethlehem Steel Trust Co. da quarantasei dollari salgono a cento l'8 di aprile; balzano a duecento il 22 luglio, valicano due settimane dopo, nell'agosto, i trecento dollari; ed inseguono irresistibili l'eccelso livello del domani.

Chi aveva nel gennaio scorso, con dieci azioni della Bethlehem Steel Trust Co. un patrimonio miserabile di quattrocento sessanta dollari, oggi vive di rendita se del poco s'accontenti: ha nelle stesse dieci azioni tremila dollari di capitale.

A chi li deve se non alla guerra?

Le azioni dell' American Can Co. che valevano quarantacinque dollari tre settimane addietro sono oggi a sessantadue; la produzione del rame, una industria in fallimento nel nostro paese, ha ritrovato sotto la stretta della guerra il coraggio, il vigore, la fortuna: i profitti della Ray Consolidated Co. che erano zero su' primi dell'anno salivano nel marzo a 741,000 dollari ed alla fine del secondo trimestre, nel giugno scorso, si realizzavano in 1,340,000 dollari!

Fino a ieri, con tutte le nostre risorse, con le nostre miniere, le nostre ferrovie, la ricchezza smisurata del nostro paese non eravamo costretti a togliere denaro a prestito sui mercati finanziarii d'Europa accollandoci un debito d'oltre mezza dozzina di miliardi pei quali paghiamo ogni anno duecento venticinque milioni d'interessi? E' scoppiata la guerra e noi abbiamo dato denari in prestito a tutte le nazioni; il vecchio debito è estinto, ed i nostri debitori pagheranno quind' innanzi a noi quello che fin qui ci è toccato sborsare a loro.

Le nostre esportazioni che arrivavano ieri alla cifra compassionevole di 807 milioni, s'approssimano oggidì ai tre miliardi, e se dobbiamo giudicare dalle cifre più recenti dateci dal Bureau of Foreign and Domestic Commerce di Washington, **cinquanta milioni di dollari** pel solo mese del luglio testè scorso, nessun paese del mondo, in nessun periodo più avventurato della sua prosperità, ha mai attinto il vertice di ricchezza e di grandezza donde irradia la repubblica tanta [illegible]

— Grondan sangue i vostri forzieri, ed è bieca d'usure adunche la gloria della nazione.

⚜

Compatisce la gente seria, pietosamente: "la miseria ed il sangue del prossimo! Mettiamo da banda un minuto, i convenzionalismi superati od ipocriti: del nostro prossimo nessuno fa conto se non per spogliarlo. Su le bilancie della nostra provvida morale, il suo sudore, il suo sangue non pesano se non in quanto ci appaiano il necessario ricostituente della vita, della forza, dell'ordine, del dominio nostro, ieri che la guerra non c'era, domani e sempre, finchè esso non ritroverà nella coscienza sua, sapientemente, provvidamente farcita di tutte le devozioni, l'improvvisa rivelazione della sua irresistibile forza.

— Non era ieri, quando la pace benediva ogni terra, così nudo, così affamato, così vile come oggi?

E del suo sangue, delle sue carni non ci siamo abbeverati e nutriti ieri, allorquando era tutto un idillio la terra, su tutti i solchi, a Chalons od a Berra, a Cherry Valley od a Ludlow, a Berlino o a Buenos Ayres, a Mosca od a Pretoria, per la salvezza di quell'ordine cui siete pur voi, piissima gente che crede in dio e si confessa nella sua parola, palladio e scudo?

Così come negli abissi della miniera o nell'androne avvelenato de la fabbrica noi suggiamo, stilla a stilla, dalle vene pallide dei suoi mocciosi, dal seno esausto de le sue femmine, dalla cervice prona di tutti gli schiavi rassegnati, il sangue del corpo ed il sangue dell'anima a colorir le guancie delle nostre donne, a costellar un sorriso su le labbra dei nostri bambini ad irradiare l'aurora della giornata fugace, a tingere la porpora del sovrano dominio in cui, venerando, consentiti devoti?

La guerra, mischia bestiale d'armi e d'armati o trama subdola d'insidie esperte, è l'aspetto costante ed immutato della vita finchè nelle nostre mani, cinto dalle triplici ritorte del ferro, della fede e della viltà, è ligio vassallo il prossimo nostro.

Non v'impietosite sul vinto, sulo schiavo incatenato e ferito; potrebbe sotto le ceneri del rassegnato torpore ritrovare la memoria e le faville d'un diritto che l'oblio millennario non è giunto a prescrivere; temprare delle sue ritorte, accendervi dei suoi belluini furori, l'ascia di Spartaco, la face dei Jacques corruschi ed implacati, far la sua guerra invece che la nostra, conquistar la sua libertà invece che la nostra fortuna, fare della terra riarsa e livellata il regno della gioia e dell'amore in luogo dell'arena selvaggia di competizioni feroci su cui si estolle il nostro imperio glo-[illegible]

Per la gloria di dio e del re, per l'estrema salute dell'ordine sociale, non sappiano le vostre labbra altro grido: **viva la guerra!**

MENTANA.

La tragedia imminente

Chi coltivi ancora un'illusione su l'indipendenza della magistratura repubblicana ed all'illusione ingenua ed ostinata chiegga la speranza di veder garantito a Davide Caplan ed a Mathew Schmidt, in omaggio alla costituzione, un giudizio imparziale e sereno nel dibattimento pubblico in cui cimenteranno in Los Angeles il 4 ottobre p. v. la vita e la libertà, può disingannarsi fin da ora.

S'incaricano di disingannarlo gli stessi giornali dell'ordine repubblicano.

La **Los Angeles Tribune** del 21 agosto ultimo, avverte infatti che "James W. Noel, il quale ha riscosso, nel processo di Indianopolis contro le organizzazioni operaie, fama particolarmente gloriosa di pubblico accusatore, è stato incaricato di sostenere l'accusa nell'imminente processo di Caplan e Schmidt imputati di assassinio in connessione cogli attentati dinamitardi che hanno mandato in rovina l'edifizio del **Los Angeles Times.**

"Noel è arrivato e dimora da parecchie settimane a Los Angeles dove ha quotidiani convegni colla procura generale e coi detectives A. H. Van Cott ed Asa Keyes particolarmente incaricati dell'istruttoria.

"La presenza di Noel a Los Angeles è stata fino a ieri un mistero gelosamente custodito...... e nessuno sa neanche oggi da chi egli abbia avuto l'incarico di sostenere l'accusa......"

La **Los Angeles Tribune** affetta un'ingenuità che mal si accorda colla sua pretesa di essere uno dei giornali più autorevoli e più sagaci della Costa del Pacifico, dal momento che non è oggi un mistero per nessuno la fonte a cui attinge il Noel la propria investitura, e le sue più doviziose risorse l'accusa.

Figure 3.3 Front page of *Cronaca Sovversiva* (Lynn, MA), 11 September 1915

Table 3.5 Anarchist periodicals' circulation, sorted by upper bound

Periodical	City	Start	End	Circulation
Il Martello	New York	1916	1946	2,500–10,500
L'Allarme	Chicago	1915	1917	2,000–6,000
L'Adunata dei Refrattari	New York	1922	1971	5,000
Cronaca Sovversiva	Barre, VT; Lynn, MA	1903	1919	3,200–5,000
L'Avvenire	Steubenville, OH; New Kensington, PA; Pittsburgh; New York	1910	1917	4,000
La Questione Sociale	Paterson, NJ	1895	1908	3,000–3,250
Eresia	New York	1928	1932	2,000–3,100
La Plebe	New Kensington, PA; Pittsburgh	1906	1909	3,000
L'Era Nuova	Paterson, NJ	1908	1917	3,000
Il Bollettino de L'Era Nuova	Paterson, NJ	1919	1919	3,000
La Jacquerie	Paterson, NJ	1919	1919	3,000
L'Emancipazione	San Francisco	1927	1932	3,000
L'Anarchia / Il Diritto / Il Refrattario	New York; Providence, RI	1918	1919	2,000
Secolo Nuovo	San Francisco	1894	1906	1,700
Domani	Brooklyn, NY	1919	1919	1,000–1,500
L'Ordine	New York	1919	1920	1,250
La Sferza	Westfield, NJ	1924	1925	1,000 ca.

Source: Data from Zimmer, 'The Whole World Is Our Country'

4,000 copies in September of the same year, when a new series edited by Malatesta was started. In December the paper stopped publishing its circulation figures, but the simultaneous decrease of its reported printing costs suggests that circulation decreased by a few hundred copies. It is probably not accidental that in the meantime two anarchist periodicals had reappeared in Italy.

How were those copies distributed, and where? We can advance some hypotheses after analysing the way the papers were financially supported. Table 3.6 provides some such data, concerning the main periodicals, as gleaned from statements in the periodicals themselves.

What becomes immediately apparent is the proportion between sales, subscriptions, and donations. Sales reach 16 per cent at best, while donations range from a minimum of 43 per cent to a maximum of 90 per cent. That

Table 3.6 Cash flow of main anarchist periodicals

		La Questione Sociale	Cronaca Sovversiva		Il Martello	L'Adunata dei Refrattari
Period		9/16/1899–4/28/1900	12/5/1903–12/10/1904	1/6/1906–8/25/1906	9/1/1923–12/15/1923	10/29/1938–12/31/1938
Weeks		32	53	33	15	9
Sales	Total	$252,92	$16,24		$364,25	$128,50
	Weekly average	$8,06	$0,31		$24,28	$14,28
	Percentage of income	16%	1%		8%	10%
Subscriptions	Total	$410,35	$475,49	$349,45	$2.239,05	$1.096,30
	Weekly average	$12,77	$8.97	$10,59	$149,27	$121,81
	Percentage of income	26%	26%	28%	49%	90%
Donations	Total	$928,10	$1.372,92	$881,06	$1.979,43	
	Weekly average	$28,87	$25,90	$26,70	$131,96	
	Percentage of income	58%	74%	72%	43%	
Income	Total	$1.597,67	$1.864,65	$1.230,51	$4.582,73	$1.224,80
	Weekly average	$49,71	$35,18	$37,29	$305.52	$136,09
Expenditure	Total	$1.631,08	$2.001,50	$1.243,90		$1.537,03
	Weekly average	50,74	$37,76	$37,69		$170,78
Balance	Total	–$33,41	–$136,85	–$13,39		–$312,23
	Weekly average	–$1.03	–$2,58	–$0,41		–$34,69

sales were limited is not surprising for periodicals that were published in small towns such as Paterson and Barre. At the same time, the fact that those periodicals thrived and enjoyed an enviable longevity in such towns means that their local readership was not their main target and that they looked at a broader readership on a national or international scale (see also Hoyt's chapter in this book).

Nevertheless, the greater volume of sales for *La Questione Sociale* suggests that the paper had a strong foothold in Paterson. The historian George Carey mentions a circulation of approximately a thousand copies in town, although it is not clear what period he refers to.[3] However, with the help of statistics about the Italian population in Paterson, it is possible to put Carey's figure in perspective. According to census data, foreign-born Italians in Paterson rose from a mere 845 in 1890 to 4,266 in 1900 and 9,317 in 1910.[4] By also including persons of Italian parentage born in the United States, the figures grow to 5,725 for 1900 and 14,748 for 1910. According to the same census data, in 1900 the average number of persons per family over the entire Paterson population was 4.5 and the number of persons per dwelling was 7.7. Thus, we can estimate that there were 740 Italian dwellings in Paterson in 1900. In 1910 the average persons per dwelling were 7.9, which yields an estimate of 1,857 Italian dwellings. Considering that *La Questione Sociale* was published from 1895 to 1908, it seems that either Carey must refer to the second half of the paper's life span or that his circulation figure of one thousand copies may be overestimated. Either way, the key conclusion is that, whatever period Carey is referring to, *La Questione Sociale* was read in a large majority of Italian households.

Still, the paper's circulation in Paterson, and probably in the United States at large, remained less than its circulation abroad. Suffice to note that around October 1899 the reported expenditure for mailing the paper abroad, which amounted to ten dollars, was five times as large as the expenditure for domestic mailing and was on a par with printing expenditure ('Amministrazione', 14 October 1899). Significantly, an issue of 18 November of the same year mentioned that the paper was mailed 'to lots of people in Italy, France, Switzerland, etc.' – a service for which a simple mention of interest was requested, rather than payment ('Ai compagni ed amici d'Europa'). In fact, the weekly administrative statements in the papers attest that sales and subscriptions mainly came from the United States. As Table 3.6 shows, these two items, together, constituted the smaller part of the income, being largely exceeded by donations, which also came for the most part from the United States. In brief, in addition to being readers, the Italian anarchists in the United States subsidised a wide distribution of the paper in Italy and other countries. In addition, an explicit confirmation that the Italian anarchists in the United States looked at Italy as their focus of interest comes from their substantial donations to the anarchist press in the homeland. Suffice to say that donations from the United States to the Ancona periodical *Volontà* (Will) in 1913–1915 amounted to 40 per cent of the total, as against 42 per cent from Italy. In particular, after the outbreak of the Great War, when the paper ran up against serious difficulties, contributions from Italy dropped to 24 per cent, while those from the United States rose to 67 per cent. Still in 1920, the donations reported over the first five issues of the newborn anarchist daily *Umanità Nova* (New humankind) were 21 per cent from Italy and 63 per cent from the United States.

The analysis of donations to the anarchist press provides an opportunity to reassess the geographical distribution of the Italian anarchists in the United States from a different angle: that of anarchists as readers and supporters of the anarchist press rather than its producers. As a starting point and a term of comparison, Table 3.7 shows the distribution of the Italian foreign-born population in 1910 and 1920 by region, according to census data.

The Italian immigrants' distribution by region remains fairly consistent from one census to the next, as well as in comparison with the distribution that we already observed for the 1890s. The comparative distribution of donations confirms what we already noticed about the distribution of the press: that the geography of Italian anarchism broadly follows that of Italian immigration, but does not mechanically reflect it, having in addition its own specific dynamics. Table 3.8 shows the distribution of the abovementioned donations from the United States to *Volontà* in 1913–1915 and to *Umanità Nova* in 1920.

The one exception to the expected order is constituted by the prominence of the South region in 1913–1915. In fact, the source of this anomaly can be pinpointed quite precisely on the map. Contributions from the South are accounted for 92 per cent by a single state, Florida, and these contributions, in turn, can entirely be ascribed to a single city, Tampa. We could zoom in even further, noting that 85 per cent of Tampa contributions came from a single area, Ybor City, the district of tobacco production, where, as already mentioned, Spanish and Italian immigrant cigar-makers were particularly combative and anarchism was highly influential. Tampa was the only US city outside of the Northeast that Malatesta, co-founder and first editor of *Volontà* in 1913, visited on his way to Cuba during his sojourn in the United

Table 3.7 Distribution of Italian foreign-born population in 1910 and 1920, by region

Region	1910		1920	
	Population	*Percentage*	*Population*	*Percentage*
Northeast	963,199	71.7%	1,163,730	72.3%
Midwest	185,066	13.8%	237,668	14.8%
West	116,707	8.7%	132,136	8.2%
South	78,153	5.8%	76,575	4.8%
Total	1,343,125	100.0%	1,610,109	100.0%

Table 3.8 Distribution of donations to Italian periodicals, by region, in Italian *liras*

Volontà (1913–15)			Umanità Nova (1920)		
Region	Amount	Percentage	Region	Amount	Percentage
Northeast	6,524.55	75.8%	Northeast	17,680.37	63.3%
South	990.85	11.5%	Midwest	8,018.85	28.7%
Midwest	926.65	10.8%	West	2,027.05	7.3%
West	108.00	1.3%	South	92.75	0.3%
Unidentified	63.10	0.7%	Unidentified	130.00	0.5%
Total	8,613.15	100.0%	**Total**	27,949.02	100.0%

States in 1899–1900, thus reinforcing a long-lasting transatlantic bond. Tampa's prominence is yet another example of the anarchist movement's own dynamics that added to the effect of immigration in shaping the geography of Italian anarchism in the United States.

Let us get into further detail by repeating the comparison between the distribution of immigrants and donations, this time by state. Tables 3.9 and 3.10 show respectively the ten states of highest Italian immigration, in 1910 and 1920, and the ten states that donated most to our two sample periodicals.

In terms of immigration, even at the level of states we still observe a strong consistency over time. The top nine states, covering respectively 84.1 and 87.0 per cent of the whole population in 1910 and 1920, are the same, in the same order, and with comparable percentages – the highest variation being New York state's decrease by 1.3 per cent. The one novelty in 1920 is the appearance of Michigan in tenth position. Indeed, between 1910 and 1920, Michigan was by far the state with the highest immigrant population growth, with an increment of 79.2 per cent, as against a national increment of 19.9 per cent.

Just as immigration statistics by state reflect those by region, so the comparison with donations by state exhibits again a common pattern with exceptions. The common pattern consists in the prominence of the four most populous states, New York, Pennsylvania, New Jersey, and Massachusetts, all in the Northeastern region. The Northeast is followed by the Midwest, primarily with Illinois but also with Ohio. Then comes the West, with California. Finally, Michigan's increase in immigrant population in 1920 is mirrored by its rise to prominence in terms of donations.

As for discrepancies between immigration and donation statistics, it is worth noting, to begin with, the role of Massachusetts, which is only fourth in terms of immigrant population but features consistently at the top of the donations list. Such prominence has different explanations in the two cases of *Volontà* and *Umanità Nova*, though. Donations to *Volontà* from Massachusetts came for 97 per cent from Lynn, where *Cronaca Sovversiva* had its offices. In turn, these were donations that *Cronaca Sovversiva* collected from all over the

Table 3.9 Distribution of Italian foreign-born population in 1910 and 1920, by state

| 1910 | | | 1920 | | |
State	Population	Percentage	State	Population	Percentage
New York	472,201	35.2%	New York	545,173	33.9%
Pennsylvania	196,122	14.6%	Pennsylvania	222,764	13.8%
New Jersey	115,446	8.6%	New Jersey	157,285	9.8%
Massachusetts	85,056	6.3%	Massachusetts	117,007	7.3%
Illinois	72,163	5.4%	Illinois	94,407	5.9%
California	63,615	4.7%	California	88,502	5.5%
Connecticut	56,954	4.2%	Connecticut	80,322	5.0%
Ohio	41,620	3.1%	Ohio	60,658	3.8%
Rhode Island	27,287	2.0%	Rhode Island	32,241	2.0%
Louisiana	20,233	1.5%	Michigan	30,216	1.9%
Other	192,428	14.3%	Other	181,534	11.3%
Total	1,343,125	100.0%	Total	1,610,109	100.0%

Table 3.10 Distribution of donations to Italian periodicals, by state, in Italian *liras*

| Volontà (1913–15) | | | Umanità Nova (1920) | | |
State	Amount	Percentage	State	Amount	Percentage
Massachusetts	2,581.35	30.0%	Massachusetts	4,539.60	16.2%
New York	2,050.35	23.8%	Pennsylvania	4,482.33	16.0%
Florida	916.00	10.6%	New York	3,862.19	13.8%
Vermont	707.65	8.2%	New Jersey	3,147.25	11.3%
New Jersey	646.25	7.5%	Illinois	2,778.95	9.9%
Pennsylvania	478.95	5.6%	Kansas	2,604.35	9.3%
Kansas	477.10	5.5%	California	1,763.00	6.3%
Illinois	272.30	3.2%	Michigan	1,500.00	5.4%
Ohio	142.25	1.7%	Ohio	835.00	3.0%
California	95.70	1.1%	Rhode Island	765.00	2.7%
Other	245.25	2.8%	Other	1,671.35	6.0%
Total	8,613.15	100.0%	Total	27,949.02	100.0%

country. As reported in its columns, the periodical collected $580.83 in favour of *Volontà* (in 1913–1915 a dollar was exchanged at a rate that ranged approximately between five and six Italian *liras*).[5] In contrast, in 1920, two thirds of all contributions from Massachusetts to *Umanità Nova* are accounted for by a single donation of 3,000 *liras* collected by Luigi Falsini in Boston – by far the largest of all donations to the periodical. The only stated source was 'a small party among comrades held on 11–10–19'. Falsini was a follower of Galleani and a member of the Gruppo Autonomo of East Boston, a very active group, forty or fifty strong, that included Nicola Sacco and Bartolomeo Vanzetti, who only months after the donation were arrested for the Braintree robbery and murder of 15 April 1920.[6] Bombs used by group members in the 'dynamite plot' of 1917–1919 – when a series of bombs were mailed or planted in several cities in response to the Red Scare anti-anarchist repression – were made in Falsini's house.[7]

Cronaca Sovversiva's role as donation collector may also help explain the absence of Connecticut – the seventh state of largest Italian-born population – from the highest-donating states in 1913–1915, since a significant amount of donations from Connecticut were funnelled through *Cronaca Sovversiva* (17 per cent of the donations collected by the periodical for *Volontà*). Indeed, the relevance of Connecticut, in step with its high immigrant presence, is confirmed by its consistent presence throughout these years among the highest-contributing states to *Cronaca Sovversiva* itself, though no obvious explanation is available as to why in 1920 its contributions to *Umanità Nova* drop again.

However, the most relevant exception to a straightforward mapping between immigrant and anarchist presence is constituted by high-donating states with a low Italian population, as these shed light on the anarchist movement's own dynamics. We have already discussed the case of Florida, centered on the anarchist (and socialist) enclave of Tampa. Lynn is another case in point, for the role of *Cronaca Sovversiva* as an anarchist network hub, as we have seen, accounts for the relevance of Massachusetts beyond its volume of Italian immigration. A similar argument can be made for Vermont, where the town of Barre, whose anarchist community we discussed earlier, accounts for 88 per cent of donations. By 1920, both Barre and Tampa had lost their relevance as anarchist enclaves, due to changing conditions, so that we no longer find the respective states among the high-donating ones. Barre's anarchist ferment began to fade prior to the First World War. The granite cutting industry's growth and mechanisation of production drove small manufacturers out and brought increased centralisation, thus undermining the anarchist emphasis on decentralized control of production.[8] The move of *Cronaca Sovversiva*'s offices to Lynn was probably both a reflection and a contributing factor of this changing situation. In Ybor City radicals paid a heavy price for their opposition to the war. The Red Scare scarred the city's radical community, with agents seizing presses and many radicals being deported.[9]

The other state of low immigration that we find among the highest-contributing states to both *Volontà* and *Umanità Nova* is Kansas. All Kansas donations come from a cluster of coal mining towns (Arma, Cherokee, Franklin, Frontenac, Girard, Mulberry, Pittsburgh) enclosed in a tiny rectangle of ten by twenty-five kilometres in Crawford County, along the border with Missouri. Crawford County mining communities ranged in size from fewer than 50 individuals to over 1,000 and were 'melting pots' of native- and European-born persons. Italians were among the most represented ethnic groups.[10]

We can appreciate the distinctiveness of anarchist enclaves such as Tampa, Barre, and Crawford County by looking at donations per capita in the states that contributed most to *Volontà* and *Umanità Nova*, respectively (Table 3.11). At one end we have states of high immigration, for which per capita donations range within a narrow interval, regardless of the significant differences in Italian-born population. At the other end we have states with very low immigrant population and per capita donations of an altogether different order than the previous groups. The sharp discontinuity between the two groups clearly shows that we are in the presence of two qualitatively different phenomena. The gap is even more striking in the case of *Umanità Nova*, where the difference in donations per capita between the first and ninth states – New York and Michigan, respectively – is seventeen times smaller than the difference between the latter and the tenth state, Kansas.

Overall, the tendency to concentration in key towns that we observed for the pre-war production of the anarchist press is confirmed by the distribution of anarchist donations in 1913–1915. The first five highest-contributing towns (Lynn, New York, Tampa, Barre, and Paterson, in this order) account for

Table 3.11 Per capita donations to Italian periodicals, by state, in ascending order

Volontà (1913–15)			Umanità Nova (1920)		
State	Population	Per capita donations	State	Population	Per capita donations
California	63,615	0.15	New York	545,173	0.71
Pennsylvania	196,122	0.24	Ohio	60,658	1.38
Ohio	41,620	0.34	California	88,502	1.99
Illinois	72,163	0.38	New Jersey	157,285	2.00
New York	472,201	0.43	Pennsylvania	222,764	2.01
New Jersey	115,446	0.56	Rhode Island	32,241	2.37
Massachu-setts	85,056	3.03	Illinois	94,407	2.94
Kansas	3,520	13.55	Massachu-setts	117,007	3.88
Vermont	4,594	15.40	Michigan	30,216	4.96
Florida	4,538	20.19	Kansas	3,355	77.63

75 per cent of all donations. In contrast, a greater geographical dispersion is observed in 1920: the first five highest-contributing towns (Boston, Paterson, Franklin in Kansas, New York, and Erie in Pennsylvania) only cover 41 per cent of all donations. This greater distribution may be partly due to the rise of new generations of anarchists in different areas from the traditional Italian anarchist enclaves. On this note, Kenyon Zimmer notes that Italian American syndicalism was predicated on industrial unionism, and therefore the map of Italian anarchism may have been partly redefined along the lines of the map of industrial unionism.[11] Moreover, Zimmer argues that 'the vast majority of foreign-born anarchists were in fact not yet anarchists at the time of their arrival ... Adoption of anarchist ideology and participation in anarchist organisations and activities resulted from their experiences as immigrant workers in America.'[12]

Be that as it may, the persistent ties of Italian-born anarchists with their homeland are all the more remarkable. The transnational character of Italian anarchism in the United States is well epitomised by the Midwest. The relative importance of the Midwest in contributing to the anarchist press in Italy grew significantly from the pre-war to the post-war years. While the incidence of its Italian-born population remained practically constant between 1910 and 1920 (ranging from 13.8 to 14.8 per cent of the entire Italian-born population in the United States) the relative weight of its support to the Italian anarchist press rose from 10.8 per cent of all donations to *Volontà* to 28.7 per cent of those to *Umanità Nova*. One could hardly associate a place as remote, isolated, and rugged as a Kansas coal mining camp with anarchism in Italy. And yet the only anarchist daily newspaper ever in Italy – if we except occasional aborted attempts – received from the American Midwest contributions that amounted to 87 per cent of what the entire movement in Italy contributed, at a time when class struggle was at a peak in Italy. By the volume of their contributions, the Italian workers of Illinois, Kansas, Michigan, and Ohio bear witness to their attachment to both anarchism and their homeland, as well as to the connectedness of the transatlantic anarchist network. Expunging those workers from the history of Italian anarchism would amount to renewing the first injustice of forcing them out of their country.

Transnational editors

The mobility of Italian anarchism between the two sides of the Atlantic Ocean was mobility of ideas, of money, but above all of people. If one were to list the chief figures of anarchism in Italy in its first half a century of life, the first names that would come to mind are probably those of Errico Malatesta, Pietro Gori, Luigi Galleani, and Francesco Saverio Merlino. Now, at different times and for periods of various length, all these militants were editors of periodicals in the United States: Merlino established *Il Grido degli Oppressi* (The cry of the oppressed) in New York in 1892; Gori contributed to *La Questione Sociale* in its first year of life, in 1895; Malatesta effected an important shift in the paper's direction, starting a new series in 1899; and Galleani was the life and soul of *Cronaca Sovversiva* for many years. As stated earlier, such mobility cannot merely be explained as a direct consequence of migration, but followed rather its own dynamics. It was mobility in both directions. Among the abovementioned four anarchists, only Galleani settled permanently in the United States, and eventually he, too, returned to Italy. Our survey of 54 names of directors, editors, managers, and publishers of Italian anarchist periodicals in the United States led to the identification of the place of death for 22 of them, associated with 29 periodicals. It turned out that 45 per cent of them died outside of the United States (eight in Italy and two in other European countries as antifascist exiles).

The editing of *La Questione Sociale* best illustrates the model of cooperation, integration, and division of labour that characterised the relationship between Italian anarchists on the two sides of the Atlantic Ocean and that somehow rationalised dynamics forced upon them. Exile and repression in the homeland on the one hand necessitated a flow of resources from the United States to Italy and on the other hand generated a flow of militants in the opposite direction. *La Questione Sociale* was established by the 'Diritto all'Esistenza' (Right to exist) group and always preserved its character of collective undertaking by grassroots militants. At the same time, a steady flow of Italian anarchists from Europe – among which the aforementioned Pietro Gori was one of the first – gave rise to a sort of alternation that ensured a sustained editing activity of high quality. One of these editors was Giuseppe Ciancabilla, who arrived in October 1898 and impressed his anti-organisationist mark on the paper, raising discontent among some group members. The controversy was resolved by Malatesta's arrival from Europe. In September 1899 Malatesta replaced Ciancabilla at the helm of the periodical and started a new series, restoring its original organisationist direction. What prompted a figure such as Malatesta to leave the social struggle in Europe to contend with an anti-organisationist for the editing of a periodical in New Jersey? The answer is simple and sheds light not only on Malatesta's motives, but also on the role of the Italian anarchist press in the United States in times of repression in Italy. At that time, a year after the 1898 bread riots and the ensuing repression, there was no anarchist periodical in Italy. Moreover, *La Questione Sociale* was one of only two Italian anarchist periodicals worldwide, the other one being in Argentina. It is plain to see why Malatesta was concerned that one of the two surviving voices of Italian anarchism had taken an anti-organisationist turn. Other editors followed one another after Malatesta's departure, including Luigi Galleani, until the paper ceased publication once it was denied mail privileges, only to restart immediately under the new title *L'Era Nuova*.[13]

What did the Italian anarchist press in the United States write about? Most periodicals shared two features: they were propaganda periodicals and were published weekly or fortnightly. This meant that many articles had no direct link to current events or, at any rate, they aimed more at commenting than informing on those events. Commentaries concerned as much Italy as the United States. For example, in November to December 1904, for four consecutive issues *Cronaca Sovversiva* devoted its full front page to a serialized long discussion titled 'Gli ultimi avvenimenti d'Italia' (The latest events of Italy), concerning a general strike. Such interest in Italy was not an interest of mere observers, but an interest of active participants in the social struggle in the homeland. Thus, in May 1899 an issue of *La Questione Sociale* contained, as a supplement, a manifesto addressed directly to the Italian people ('Al Popolo Italiano'), signed by a 'Comitato Rivoluzionario degli Stati Uniti' (Revolutionary committee of the United States). A year after, an Italian anarchist from Paterson actually stormed into Italian history in the most sensational way, by killing King Humbert I. So it was that an obscure New Jersey town got overnight its worldwide reputation as an 'anarchist nest' and a small anarchist newspaper with a circulation of 3,000 copies was graced with a full-page feature article in Joseph Pulitzer's *New York World*, in which the periodical's offices, masthead, and current editor were portrayed under the eight-column title 'Malatesta, the king-destroyer, leader of the Paterson anarchists' (Figure 3.4).As a member of the editing group recalled, this was also the time when the periodical's circulation peaked.[14] Unsurprisingly, the involvement of the US Italian anarchist press in the events of Italy could only grow after Fascism's rise to power, when no opposition press was tolerated on the national territory. The antifascist struggle was fought as much directly, in the various Italian colonies, as through the press. Newspapers such as Carlo Tresca's *Il Martello* made that struggle their main field of action.

Conclusion: the other Italy

In conclusion, the Italian anarchist press of the United States provides evidence that its producers and readers considered themselves and their comrades in Italy as a single movement that stretched across the two sides of the Atlantic Ocean – and we, too, must consider them as such. Theirs was a transnational movement that crossed the territorial boundaries of Italy but at the same time preserved a national identity. What could such identity lie in, for an anarchist? Despite all attempts to define it, the concept of 'nation' is notoriously elusive. As one of the most authoritative scholars in this area, Ernest Gellner, has noted, both 'cultural' theories à la Herder and 'voluntaristic' theories à la Renan ('the existence of a nation is a daily plebiscite') emphasise relevant facets, 'but, just as obviously, neither is remotely adequate'. For Gellner, it must be acknowledged that another agent or catalyst is crucial for group formation: 'fear, coercion, compulsion'.[15] This is certainly true if we presuppose, as Gellner and all theorists of nationalism do, that the creation of a state is the necessary completion of a nation's formation process – that is, if we think of a nation as a nation-state. If that were the case, the anarchists' national identity would be *ipso facto* a contradiction in terms. What

Figure 3.4 New York World of 8 January 1900

happens, though, if we drop that hyphen that joins the 'nation' and the 'state' in indissoluble matrimony? Does the concept of 'nation' become vacuous or self-contradictory? It does not (see also Ferretti's chapter in this book). Rather, it ceases to be the exclusivist concept peculiar to nationalism, which demands unconditional and undivided allegiance to a single nation and presupposes borders watched by the respective police corps, to become a concept that celebrates diversity, plurality, overlapping, and coexistence, and presupposes spontaneous processes of territorial, linguistic, ethnic, and historic assimilation and differentiation. This was the anarchists' concept of nation. It was the heir of the universalism that characterized Mazzini's 'Europe of nations', in which sovereign nations were to coexist in a spirit of mutual respect and harmony.[16]

Thus, in an article published in London, which both *Cronaca Sovversiva* and *L'Era Nuova* hastened to republish on the same day, 11 May 1912, Malatesta explained that 'love of birthplace, or rather the greater love of the place where we were raised ... preference for the language we understand and the consequent closer relationships with its speakers are natural and beneficial things'. For this very reason, though, in the same article the cosmopolitan Malatesta declared that when 'Italy invades another country and Victor

Emmanuel's infamous gallows is erected in Tripoli's market square and kills, the Arabs' revolt against the Italian tyrant is noble and blessed', just as noble and blessed was the Italians' revolt against the Austrian tyrant.[17]

Taking up the subject of 'love of one's country' again in 1921, Malatesta declared that it had been an error to 'let the conservatives and the petty tools of the bourgeoisie monopolise somehow the war cry 'long live Italy' and thus manage to persuade the simpletons that we wish ill upon the country where we live'.[18] I hope this survey of the Italian anarchist press in the United States helps rectifying the misapprehension that Malatesta lamented. A century after the carnage of the First World War and in contrast with the national rhetoric still associated with it, it is sobering to remind ourselves of another Italy, which nurtured a deep sense of national identity without nationalism and an intense love of the country without patriotism.

Notes

1 Lord, Trenor, and Barrows, *The Italian in America*, 4–6. We have rearranged the data, which were organised by divisions, so as to adapt them to the current partition in regions as defined by the United States Census Bureau.
2 Malatesta, *Verso l'anarchia: Malatesta in America, 1899–1900*, 60.
3 Carey, 'La Questione Sociale: An Anarchist Newspaper in Paterson, N.J. (1895–1908)', 291–292.
4 All historical data about United States population cited hereafter are gleaned from various U.S. Census Bureau sources available online at http://www.census.gov/libra ry/publications.html and particularly the *Statistical Abstract Series*, available at http://www.census.gov/library/publications/time-series/statistical_abstracts.html, and publications for specific years, available at webpages of the form http://www.census. gov/library/publications. *yyyy*.html, where *yyyy* stands for the year in question. All webpages were accessed between 13 and 17 July 2016.
5 I would like to thank Andrew Hoyt for kindly giving me access to his extensive database of subscriptions to *Cronaca Sovversiva*.
6 Avrich, *Anarchist Voices: An Oral History of Anarchism in America*, n. 227.
7 Avrich, *Sacco and Vanzetti. The Anarchist Background*, 157–158.
8 Woodsmoke Productions and Vermont Historical Society, 'Anarchist Movement in Barre', *The Green Mountain Chronicles* radio broadcast and background information, original broadcast 1988–1989, accessed on 27 August 2016 at http://vermonthistory. org/research/research-resources-online/green-mountain-chronicles/the-anarchist-movement-in-barre-1920.
9 Mormino and Pozzetta, 'The Radical World of Ybor City, Florida', in Cannistraro and Meyer (eds), *The Lost World of Italian American Radicalism: Politics, Labor, and Culture*, 259.
10 Powell, 'Former Mining Communities of the Cherokee-Crawford Coal Field of Southeastern Kansas'.
11 Zimmer, *'The Whole World Is Our Country': Immigration and Anarchism in the United States, 1885–1940*, 17.
12 Zimmer, *'The Whole World Is Our Country'*, 18.
13 On Malatesta, *La Questione Sociale*, and transatlantic cooperation, see also my *Making Sense of Anarchism: Errico Malatesta's Experiments with Revolution, 1889–1900*, ch. 8.
14 N. Cuneo, 'Vent'anni!', *L'Era Nuova*, 17 July 1915.

15 Gellner, *Nations and Nationalism*, 53–54.
16 For a fuller discussion of this subject see my 'Nations without Borders: Anarchists and National Identity', in Bantman and Altena (eds) *Reassessing the Transnational Turn: Scales of Analysis in Anarchist and Syndicalist Studies*.
17 Malatesta, 'La guerra e gli anarchici' (War and the anarchist), in Turcato (ed.), *'Lo sciopero armato': Il lungo esilio londinese, 1900–1913*, 246–247.
18 'L'amor di patria' (Love of one's country), *Umanità Nova* (Rome), 24 August 1921.

4 Humour, violence and cruelty in late nineteenth-century anarchist culture

Julian Brigstocke

In this chapter I explore the ways in which late nineteenth-century French anarchism created a link with forms of experimental and subversive humour that has remained an important part of the movement until the present day. Focusing on the culture of humour in anarchist cultural spaces in Paris of the 1880s and 1890s, I explore the motivations for the conjunction of humour and anarchism, and examine some of the political reverberations of this practice in the context of the rise of anarchist violence or 'propaganda by the deed' during the 1890s.

In the 1880s, anarchists attempted to establish the working-class area of Montmartre, on the periphery of Paris, as a utopian space of creativity, hedonism, moral experimentation, and joyful humour. A number of cabarets and cafes, some politically neutral but anarchistic in outlook, and others explicitly anarchist, flowered in the area and revelled in a culture of excessive, carnivalesque humour. By the end of this decade, however, as the doctrine of anarchist violence known as 'propaganda by the deed' became established, key figures of Montmartre humour became transfigured and transformed. This chapter will explore how the ambivalence and ambiguity of humour gave way to a search for new forms of transcendent morality.

Anti-authoritarian humour

Anarchist political culture has a long tradition of engaging with humorous, carnivalesque forms of protest and political practice. This tradition dates back to the early history of anarchism, where anarchists became interested in the creative possibilities of humour for challenging the authority of dominant discourses concerning the life and vitality of the city.[1] A significant amount of critical attention has recently been devoted to these playful, affirmative forms of anarchist humour. Sharpe et al. analyse the humorous forms of resistance embodied by 'culture jammers' such as the Yes Men, and the effects of their practice of using the internet to 'borrow' the identity and authority of authoritative institutions such as the World Trade Organization.[2] By avoiding didacticism, and avoiding the logic of representation, such protests create events that prompt thinking towards something new and original. Similarly,

Paul Routledge has shown how groups such as the Clandestine Insurgent
Rebel Clown Army play an important role in confusing hierarchically
imposed categories, undermining authority by holding it up to ridicule, and
generating 'sensuous solidarities' and 'complex, contradictory and emotive
co-performances and resonances with police, other protestors and the
public'.[3] This resonates with Kanngieser's analysis of the role of humour in
creating 'transversal' experiments with making new worlds.[4] Such experiments
create links across different terrains of subjectivities, positions, expertise, and
worlds. Transversal forms of world-making encompass qualities of mobility
(traversing domains, levels and dimensions), creativity (productivity, adven-
turousness, aspiration), and self-engendering (autoproduction, self-positing).[5]
As Kanngieser argues, humorous transversal performances have the capacity
to re-appropriate hierarchical authority structures by redistributing forms of
expertise, and in doing so they can challenge hierarchies of value.

According to the philosopher Simon Critchley, humour is an important
tool for a pacifist anarchist politics based on a fidelity to an 'unfulfillable
demand', a demand that the subject acknowledges but can never fully meet.[6]
Critchley argues that humour works through displacement, generating something new by defying our expectations. Humour opens up a fracture in the
order of things, generating a difference between how the world is and how it
might be. According to Critchley, humour recalls us to the modesty and limitedness of the human condition, a limitedness 'that calls not for tragic-heroic
affirmation but comic acknowledgement, not Promethean authenticity but
laughable inauthenticity'. This account of humour is an important move in
Critchley's argument for a neo-anarchist politics based on infinite responsibility
arising in relation to situations of injustice. He draws on the pervasive use of
humour in contemporary anarchism's 'new language of civil disobedience',
which 'combines street-theatre, festival, performance art and what might be
described as forms of non-violent warfare'. These comical tactics, he argues,
exemplify the effective forging of horizontal chains of equivalence or collective
will formation across diverse and otherwise conflicting protest groups.

What these insightful accounts of anarchist humour fail to capture, however,
is the complex relationship between humour and violence. I would suggest that
there are often forms of violence embedded within even wholly pacifist practices.
Moreover, the relationship between humour and violence becomes even more
pressing in the historical context of fin-de-siècle Montmartre, where physical
violence (whether the violence of the state or of revolutionary violence) was
never far from the surface, and where the dominant style of macabre humour
made continual reference to various forms of violence such as murders,
robberies, uprisings, executions, suicide, and forced labour.

Indeed, whilst it is tempting to associate humour with subversion and
resistance, a number of critics have demonstrated how easily humour can be
allied with exclusion, hatred and violence. Jokes can act as powerful affective
vehicles for creating and consolidating boundaries: securing, in other words,
the integrity of the inside and outside of social groupings. Humour has the

power both to include and to exclude. There are numerous ways in which humour has the power to include and exclude, since in order to work it relies upon nuanced sensitivities to shared histories, traditions or codes.[7] Philosophers such as Jean-Paul Sartre, Theodor Adorno and Boris Groys, for example, have written about the important role of humour in anti-Semitism, Fascism, and Stalinism. Humour is a socially sanctioned means of breaking taboos and 'saying the unsayable', which can be useful for the articulation of extremist views. As Michael Billig has shown, racist humour takes advantage of humour's ability to transgress limits by articulating racist ideas that are normally taboo, and claiming that they are 'just a joke'.[8] Because having a 'sense of humour' carries strong normative force in modern societies, it is much easier to get away with offensive speech when cloaking it in humour. This issue was forcefully dramatized in recent debates around the use of humour as a form of symbolic violence against Muslims through derogatory portrayals of Mohammed.[9] Humour, in such cases, is deployed in the services of hatred, resentment, and violence.

Rather than viewing the politics of humour in terms of a simplistic binary of violence/resistance, however, I would like to draw on the thought of Gilles Deleuze to consider the ways in which humour engages with the everyday violences of law, signification and subjectification. In Deleuze's writings, we find an account of humour that explores humour's essential link with cruelty and violence, and recognises the creative potential of humour's 'cruelty'. In works such as *The Logic of Sense* and *Coldness and Cruelty*, Deleuze draws an important distinction between irony and humour. Whereas irony refers back to the power of reason and the sovereignty of the subject, humour is the art of thinking the noises, sensations, affects and sensible singularities from which bodies are composed.[10] Humour contains pulsations of life that have the capacity to disrupt concepts and categories. Humour is an art of singularities, of events that are not meaningful, or structured according to a logic of before and after. Radical forms of humour dissolve high–low distinctions, and instead work through a play of surfaces. Humour, moreover, is essentially corporeal: it takes the human subject down to its corporeal origins, so that the self does not appear as an organising subject but as an assemblage of incongruous body parts.

Deleuze's account of humour draws together a Nietzschean account of the roots of law and morality in violence, a Bergsonist account of humour as a way of affirming the vitality, suppleness and flexibility of life by taking enjoyment in the forms of violence where living humans are momentarily reduced to mechanistic, lifeless objects, and Artaud's notion of a theatre of cruelty. For Deleuze, morality becomes possible when cruelty is given a meaning and becomes subject to judgement: the forceful interactions of bodies are made subject to law. Humanistic and moralistic points of view enable cruelty, violence and force to be seen as exercises of law or punishment. Humour, however, provides a means of making visible the cruelty of life, for example through the meaningless enjoyment of another's suffering. In slapstick humour, the viewer takes delight in watching a body fall, in misfortune, at the collapse of law, at misguided logic, at the exposing of hypocrisy. Humour's violence does not justify itself through reference to a higher morality, but assaults conscious intentions and determinate meanings, and remains situated at the level of the body. Humour is a disruptive force, and hence a form of cruelty that does not conceal itself through reference to hierarchical laws and norms. As a form of cruelty, humour undermines vertical hierarchies and the authoritarian morality of judgement and law.

Viewing humour in this way, our analytical eye refocuses away from judgements between violent and non-violent humour, to distinguishing between forms of humour that refer to a transcendent morality or subjectivity (thereby contributing to an implicit, veiled violence), and forms of humour that remain on the surface, violently subverting hierarchical categories, meanings, or norms. These latter forms of humour make visible life's cruelty: the violence (disrupting forms, evading categories, provoking bodily affects) that is a necessary accompaniment to vital creativity.

An anti-authoritarian utopia

During the 1880s, Montmartre, a working-class district on the outskirts of Paris, started to establish itself as a centre of anti-authoritarian counter-culture. In the aftermath of the Franco-Prussian war and the Paris Commune, the tentative and fragile Third Republic put a renewed emphasis on a return to order and authority, symbolised by the vast new Basilica de Sacré Coeur built on Montmartre (on a site that was symbolic of the start of the Paris Commune uprising) and visible throughout the city. If Montmartre is now celebrated by the tourist industry as a quirky suburb of artistic creativity, this is an image that is largely divested of key elements that first made Montmartre both notorious and alluring during the early 1800s: sex, alcohol and violence.

A bawdy 1882 poem in *Le Chat Noir* satirical literary journal, written in a faux-mediaeval French (in a typically Montmartrois appeal to the non-modern), captured something of the reputation that Montmartre was beginning to acquire as an outpost for the marginalised, the down-at-heel, and the immoral – including prostitutes and their clients.

> Ragged philosophers of filth,
> Loose women with sickly looks,
> Trollops, cheats, strollers, whores
> Who honour our pillows,
> Endure our hands to caress
> Your gelatine breasts,
> Cleavage in distress,
> Under the grove of the Assassins.
>
> Students, robbers, cheerful sorts,
> Flunkeys, cabbies, little saints

In prison uniforms and shackles,
Captains, fighters, swordsmen,
Suffering from treacherous thirst,
Fly here in swarms
For Communion at the dive Mass
At the altar of the Assassins.

Poets gnawing on bones,
Toothier than young boar,
Can fill up here with delicate morsels
From clay basins.
Pipers of jubilant harmonies,
Tickling the ivories
– Come here to pick the flowers of drunkenness
While blessing the Assassins.

City-dwelling lovers of the gutter,
Come here, far from the cops,
Deliver your joy to the arbours,
Under the gaze of the Assassins.[11]

The poem captures some key qualities of Montmartre's allure: a decadent embrace of hedonism, immorality and creativity; a home for the down-at heel ('Come…city-dwelling lovers of the gutter'); a promise of freedom from the police ('Come here, far from the cops'); and a gently humorous realisation that the threat of violence implied by the talk of assassins merely refers to the 'Cabaret des Assassins' in Montmartre, another iconic site of Parisian counter-modernity. As we will see, more explicitly anarchist cabarets in Montmartre also played on this playful conjunction of humour and the threat and memory of violence.

Central to Montmartre's cultural experiments was a focus on undermining the solemnity of the new Third Republic through ironic and pantomimic humour, within cultural spaces that privileged minor, fragmentary art forms. Cultural performances in Montmartre routinely joked about establishing Montmartre as an independent, autonomous, kingdom, far superior in its modern attitudes to the drab bourgeois city below it. Soon Montmartre's anti-establishment values and outlandish forms of humour, and its role as a symbol of the Paris Commune and its brutal repression, led Montmartre to become home to a number of cafes and cabarets that were either explicitly anarchist or exhibited anarchist values.[12] These cafes and cabarets explored modern issues through humorous performances, songs, minor artforms such as shadow theatre and puppetry, and political discussion. Anarchists were attracted by the neighbourhood's reputation for autonomy, rebelliousness, and experimental morals and forms of life.[13]

Above all, anarchists were attracted by Montmartre culture's carnivalesque questioning of authority, in particular its reaction against the renewed emphasis during the 1870s on traditional morality and religious observance.[14]

The artists, bohemians and political radicals who congregated in Montmartre viewed it as a semi-utopian space in which new forms of morality, politics, and cultural expression could be practised, free from outside interference. Anarchists saw Montmartre as an alternative society that valued creativity, non-conformism, art, and hedonism. Montmartre came to symbolize an anarchist version of utopia, celebrating free creativity, local autonomy, and a balance of urban and rural elements. Montmartre 'preserved its own sacred space from which to gaze down upon the metropolis, countering its economic dependence with cultural autonomy and radicalism'.[15] Despite its subjection to the rest of Paris and the destructive forces of capital, Montmartre was viewed as retaining a certain distance and autonomy from the capital. Its emphasis on harnessing the vitality, playfulness and anti-authoritarian dynamics of laughter made Montmartre an obvious home in which anarchists could congregate. Its culture was characterized by a mixture of humorous joy (one of the slogans of Montmartre was 'Mont joie [Mount Joy] – Montmartre!') and melancholia at the violence of capitalist modernity. The result of this authoritarianism was that true creativity had to hide, disguised as a naïve clown, in the poor and marginalized suburbs of the city. Thus the figure of the suicidal clown Pierrot became one of the symbols of Montmartre.[16] In a huge 5x3m painting by Adolphe Willette, which covered one wall of the Chat Noir cabaret and depicted a carnivalesque torrent of bodies pouring down from Montmartre into Paris, a suicidal Pierrot leads the procession. Following him a young woman rides on the back of an enormous black cat. A skull supervises the procession in place of the Moon, and angels dance in the sky. Montmartre seems to be poised at the intersection of the living and the dead. The painting captures the combination of excessive energy and macabre pessimism of the spirit of Montmartre.

Although there is a long tradition of the 'sad clown' in Romanticism, Willette was the first to depict Pierrot as a bohemian artist (and, conversely, the bohemian artist as Pierrot).[17] In doing so, Pierrot's iconography changed: he became noticeably paler in face and started to dress in black. His pallor, resembling that of moonlight, emphasized that he was an outcast, a child of the Moon. His black clothes were a parody of those of the bourgeois class: he was to present the bourgeoisie at once with a degrading parody of itself and an image of its victim. The poet Gustave Kahn described Willette himself as a real life Pierrot character:

> Willette is … the very pavement of the Paris streets, come alive with all its *blague*, all its wit, lit up by tenderness, giving off smoky glimmers of passing political passions. Willette is patriotic, Willette is working-class. He will be generous, he will be cruel, he will be sympathetic, he will be hateful, according to the direction of the wind … There is in Willette a figure who will man the barricades for the fun of it.[18]

This joyful utopianism was expressed in the humorous campaign, conducted over several years, for Montmartre to gain freedom from the rest of Paris city.

Standing as a candidate in the municipal elections in 1884, the patron of the Chat Noir cabaret Rodolphe Salis proclaimed:

> What is Montmartre? Nothing! What should it be? Everything! The day has finally come where Montmartre can and must assert its right to autonomy against the rest of Paris. Indeed, in its association with what is commonly known as the capital, Montmartre has nothing to gain but the burden of its humiliation. Montmartre is rich enough in wealth, art and spirit to lead its own life ... Montmartre deserves to be more than a district. It must be a free and proud city.[19]

This tongue-in-cheek autonomist movement echoed the demands of the Paris Commune for municipal autonomy, and also parodied the language of Abbé Sieyès' 1789 revolutionary pamphlet 'What is the Third Estate?', which asked three questions concerning the third estate (the common people): 'What is the Third Estate? Everything. What has it been until now in the political order? Nothing. What does it ask? To become something.' An accompanying article in *Le Chat Noir* continued this emphasis on revolutionary autonomy:

> Montmartre is isolated because it is self-sufficient. This centre is absolutely autonomous. It is said that in a small village located a long way from Montmartre, which travellers call Paris, a local academy is discussing the conditions of municipal autonomy. It is a long time since this question was resolved in Montmartre.[20]

This celebration of a utopian city, divorced from and looking down haughtily on the corrupt city below, was the predominant spatial motif of Montmartre counter-culture. Montmartre was portrayed as a city that was full of life, energy and vitality, and contrasted with a Paris that was lifeless and soulless. Toulouse Lautrec's illustration (Figure 4.1) for a short-lived anarchist journal called the *Vache Enragée*, for example, portrayed an angry bull chasing a terrified bourgeois from Montmartre to Paris, with two clowns following on a bicycle (itself a potent symbol of fin-de-siècle modernity). The angry cow (vache enragée) was, in fin-de-siècle popular culture, a metaphor for hunger, poverty and abjection. It stood in opposition to the 'veau d'or', the golden calf which symbolised greedy capitalism. The image bristles with an energy that draws together humour and violence: the naïve, capricious clowns – symbols of Montmartre popular and avant-garde culture – pursuing the bourgeois invader with the uncontrolled anger of the enraged beast. Montmartre is imagined as space of vital humour that will fiercely defend its autonomy and independence.

In the humour of Montmartre's 1880s bohemian counter-culture, Montmartre emerges as a space of both creativity and threatened violence, whether Pierrot's suicide or his revolutionary violence against the bourgeoisie of Paris. Humour not only kept memories of the Commune alive (in a political and artistic

Figure 4.1 Henri de Toulouse Lautrec, Cover for the journal *La Vache enragée*, 1896, Paris

culture that was trying hard, as Collette Wilson has shown, to erase its traces), but evoked the possibility of new forms of community that could gain some autonomy from Paris, albeit at the price of poverty, marginalisation, and the imperative to disguise itself in the figure of the naïve, affirmative, yet suicidal clown.[21] Humour played the function of both making the violence of modernity visible, but also evoking the possibility of revolutionary violence in ways that would not immediately result in censorship and arrest.

Cabarets such as the Chat Noir, whilst demonstrating a broadly anarchistic outlook, were not directly affiliated to the anarchist movement, and at least claimed to remain apolitical. As the movement expanded, however, a number of explicitly anarchist cultural spaces emerged that used humour as a way of exploring the violence of modernity in far stronger terms.

A true fountain of manure

A typical example of the intersections between political anarchism and bohemian humour was a cabaret known as the Taverne du Bagne (Penal Colony Tavern). The cabaret, opened by the notorious revolutionary Maxime Lisbonne, and frequented by both anarchists and curious bourgeois, was a kind of theme bar that humorously mimicked the conditions of forced labour camps (the 'bagne'). Many defeated Communards had been deported to the South Pacific island of New Caledonia in 1871, but returned to Paris following an amnesty in 1880.

The Taverne du Bagne used humour to re-enact for its patrons the unpalatable truth of the Third Republic's birth in violence and civil war. Huge paintings adorned the walls of the torture and bloody martyrdom of anarchists. Bar staff dressed as forced labourers. Paul Lafargue wrote of the cabaret to Friedrich Engels: 'Lisbonne, professional ham, has had the genial idea of opening a cafe where the doors are barred, where the tables are chained, where all the waiters are dressed as galley slaves, dragging chains ... The success has been crazy.'[22] Another observer described the scene:

> A seedy light fell from the ceiling and a few dirty glass lanterns were hung on the pillars ... As soon as the door opened, you were received with a barrage of insults ... It was a true fountain of manure.... Waiters had the air of bandits and a three-day beard. They knew how to walk in their tethers with their large hooves [i.e. their ball and chain]. But from time to time they would linger on the shoulders of the lovely ladies, breathing their perfumes, offering these doves a *frisson* of the guillotine.[23]

As with many Montmartre cabarets, a satirical journal was launched to publicize the cabaret's distinct brand of anarchist humour to the general public. 'Between Paris and Montmartre', the journal exclaimed,

> the ex-convict Maxime Lisbonne has just resurrected and revived the penal colony ... Staff, condemned to serve in the tavern, have been picked from former officials, traders, industrial workers, financiers, property owners, priests, brothers, friars, who are suffering their sentences so as to live honestly ... All you who have entered the penal colony – and who, moreover, have got out – thank you for the constant kindness with which you have treated the convicts ... You have helped in the rehabilitation of the fallen, the moralization of rogues.[24]

In a society still scarred by memories of the Commune, humour of this nature was deeply shocking. The humorous popular culture of fin-de-siècle anarchism clearly participated in an emerging avant-garde sensibility towards awakening thought through shock effects. The joke here works on several levels. First, there is the incongruity and shock of offering up a dramatisation of punishment as enjoyable spectacle. The effect of this is to undermine the authority of the legal system by dramatising the nature of punishment as sadistic pleasure or revenge, rather than a genuine means of reforming criminality. Against the ideology of law as a rational procedure based on a reform of criminality to normality, the cabaret stages the authority of law as an affective system based on the pleasures of cruelty and violence. Second, there is the reversal of dominant hierarchies: it is the bourgeoisie ('officials, traders, industrial workers, financiers, property owners...') who are humorously revealed as the villains who should be in prison and in need of reform. The noble anarchists are there to help them reform. Third, there is the humour of an encounter with death: offering the bourgeois 'doves' the *frisson* of death. The anarchists who serve these ladies are liable to suffer the law's deadly violence: but criminality here is given a human face and is not allowed to be medicalised through abstract logics of pathology and degeneracy. Through humour, the violence of law – and its lack of legitimacy and natural authority – is laid bare. Similarly, the vast paintings of the execution of anarchists make visible the origins of the Third Republic in violence, and hence implicitly undermine the Republic's claims to authority and legitimacy.

Yet cultural spaces such as these remained highly ambiguous, not least because of their considerable financial success, which took advantage of the middle-class taste for 'slumming': visiting the dark and degenerate spaces they read about in highly sensationalist terms in the press. Such cabarets arguably trivialised the events, making them into a cheap diversion.[25] The cabaret's distinctive brand of anarchist humour seemed to cheapen anarchist ideals even as it derided its hated bourgeois clients. The Taverne du Bagne, for all its anarchist commitment, was a profitable space of bourgeois diversion and consumption; indeed, it was so successful that one newspaper could run with the eye-catching headline, 'All Paris in jail'.[26] Yet the form of anarchism explored here was not only a commercial gimmick. The Taverne du Bagne was a known meeting point for anarchist conspirators (as regular reports from the continual police surveillance carefully noted), and was a venue for rousing talks by the legendary Communard Louise Michel. What the success of the cabaret most strongly demonstrates is the extent to which humour can serve as an effective means of exploring and communicating radical and uncomfortable ideas to those who do not share these views, precisely *because* it is relatively innocuous and unthreatening. In spaces such as the Taverne du Bagne, bourgeois visitors could learn something about anarchist culture, and the 'underworld' more generally, without fearing for their lives.

Kings of derision

To develop a fuller theoretical understanding of fin-de-siècle anarchism's paradoxical culture of derisory humour, and its link to acts of political revolt, I wish to travel on a slight detour to Michel Foucault's remarks concerning the genealogy of nineteenth-century militant movements. Foucault links the nineteenth-century militant ideal of revolutionary life as a violent, shocking manifestation of truth to the scandalous philosophy of the Cynics.[27] Cynics practised a form of truth-telling (parrhesia) that involved living their lives as a polemical dramatisation of shocking truths. Cynicism involved a form of 'militancy in the open ... that is it say, a militancy addressed to absolutely everyone ... which resorts to harsh and drastic means, not so much in order to train people and teach them, as to shake them up and convert them, abruptly'.[28]

For the Cynics, humour was a key device in establishing the truth-teller's authority. Humour was a telling of scandalous truths to power in ways that were not overly threatening or aggressive and so were capable of maintaining productive emotional links between ruler and ruled. A Cynic such as Diogenes positioned himself as a kind of counter-king, a 'king of derision' who, 'by the existence he has chosen, and by the destitution and renunciation to which he exposes himself, deliberately hides himself as king'.[29] This emphasis on derisory humour partly explains the Cynics' habit of passing down their teachings through jokes and humorous anecdotes. This aspect of Cynic militancy is explored in more depth by Peter Sloterdijk, who emphasises the ways in which the Cynics made the truth dependent on courage, risk and 'cheekiness', contrasting this form of 'Kynicism' with the reactive cynicism of modern culture.[30] Sloterdijk discerns in Kynicism a 'pantomimic materialism' that refuted the language of the philosophers with the capers of a clown. It 'represent[ed] the popular, plebeian rejection of the official culture by means of irony and sarcasm'.[31] Most importantly, it tackled afresh the question of how to say the truth, speaking truth to power through a brute materialism, a 'dialogue of flesh and blood'. In Cynic militancy, truth is spoken through a materialist laughter that explodes with the material, vital energies of blood, urine, faeces, and sperm.

Supporting Foucault's brief suggestion about nineteenth-century militants' reactivation of elements of the Cynic spirituality of truth, we can find a clear echo of such comedic figures of counter-authority in the anarchist milieu of late nineteenth-century Montmartre. The figure of the king of derision was directly echoed, for example, in a procession organised by the Chat Noir cabaret. Rodolphe Salis, the patron of the Chat Noir, was known for greeting wealthy customers with a manner described as 'humility that was not entirely free of insolence'. He would loudly ask late-arriving customers questions such as 'When did you get out of prison?' or, to a man accompanied by his wife, 'Where's the girl you were with last night?' And in one typically Montmartrois event, Salis was crowned as the derisory King of Montmartre.

[Salis] had to wear a gold suit and incredible fabrics, and hold a sceptre. After receiving homage from the people, he went to take possession of the Moulin de la Galette. He paid a visit there, hiding his Royal clothes under an ulster coat, accompanied by painters and poets armed with halberds, who, along the whole length of the butte, and to the bewilderment of the onlookers, cried: 'Long live the King!'[32]

Such a provocation, coming only a short time after France had narrowly avoided a Monarchist restoration, was utterly scandalous, and it is only because of the excessive absurdity and humour of the event that it was tolerated by the authorities. The parade proclaimed Salis as a king of derision who would use derisory humour to successfully speak truth to power.

The derisory humour associated with anarchist cabarets, I am suggesting, can be theorised as a form of truth-telling that sought in humour a way of acquiring the authority to speak truthfully against the elite power. Humour expressed anarchists' values of freedom, creativity, and distrust of all externally imposed authority. It did so, however, in ways that retained bonds of care with those to whom they spoke. Montmartre humour, whilst raw and angry, always retained a certain warmth for those whom it mocked. Whilst this bonhomie between bohemian and bourgeois has often been seen as evidence of bohemians' basic complicity with bourgeois culture, an alternative reading would view it as a recognition of the importance of retaining affective relations of care between dominant and dominated, rather than succumbing to an oppositional, adversarial politics that falls back upon established political identities and groupings rather than allowing new, autonomous forces and subjectivities to develop.

Awakening truth

By the 1890s, as the Third Republic became increasingly stable, many militants grew frustrated with conducting politics at the level of culture and ideas, and disillusioned with the prospect of collective revolutionary action. The doctrine of 'propaganda by the deed' – communicating by acts of revolt rather than through words – came to prominence.

Propaganda by the deed attempted to awaken this spirit of revolt through a series of attacks in the bourgeois heartlands of Paris. In doing so, it made use of a distinctively modern technology: dynamite. Over the last twenty years, dynamite had become an iconic modern technology. As Sarah Cole argues, the invention of dynamite in 1866 marked an important moment in the history of modernity. Dynamite became an immediate sensation.

The violence of dynamite reverberated in every sensory register as something novel ... from its chemical smell, to its shattering sound, to its extreme tactile effects ... It shattered, exploded, ripped, and tore; ... and its employment for radical causes suggested a future with unknowable and potentially frightful contours.[33]

With the turn to dynamite came an image of anarchism that was shorn of its playful and carnivalesque elements. In popular stereotypes (perhaps most famously depicted in Joseph Conrad's novel *The Secret Agent*), the anarchist was a sombre, anti-social, over-sensitive, and incompetent loner. The Russian anarchist Sergey Nechayev's infamous pamphlet *Catechism of a Revolutionary* had provided the template for this image, describing the revolutionary as 'a doomed man' with 'no personal interests, no business affairs, no emotions, no attachments, no property, and no name'.[34] The revolutionary, he wrote, must suppress 'all the gentle and enervating sentiments of kinship, love, friendship, gratitude, and even honour'. He must be prepared to destroy anything and anyone. This striking image of the nihilistic anarchist revolutionary solidified during the period of violence in the 1890s. Anarchist culture itself contributed to this imagery in a series of writings in celebration of the most famous anarchist bomber, known as Ravachol.

The execution of Ravachol became something of a foundational event in French anarchism, and prompted a series of retaliatory attacks.[35] Ravachol had lived his life in poverty, resorting to minor crimes on occasion, and was extremely hostile to the violence of capitalism and religion. Responding to the massacre of nine protestors in a workers' demonstration in Fourmies in 1892, as well as police violence against anarchist protestors in Clichy, Ravachol planted a suitcase full of explosives in the living quarters of the Advocate General and also the public prosecutor of the Clichy affair. No-one was hurt but significant damage was done to the properties. Ravachol was soon arrested, and eventually sentenced to death for previous murders he was alleged to have committed. His vehement bravery in the face of death made him an instant martyr. As one newspaper reported Ravachol's death:

> His love of melodramatic display was exhibited in his last moments. As he was dressing to go to the guillotine, he kept singing a song, which had for its refrain, 'To be happy, hang the landlords and cut the priests in two', and this he sang in the prison-van and at the foot of the scaffold. When the assistants took hold of him, he turned round and cried out, 'I have something to say, citizens,' but he was overpowered, and held fast by the ears. In this horrible position he still had strength enough to cry, 'Vive la – ' before the knife dropped.[36]

Immediately after his death, a series of articles were published that raised Ravachol to the level of a religious martyr. Paul Adam, for example, idealised Ravachol as a modern-day Redeemer:

> Ravachol remains as a propagator of the great idea of the ancient religions which advocated a search for individual death for the Good of the world; sacrifice of the self, of his life, and his fame for the exaltation of the poor, of the humble. He is definitively the Restorer of Essential Sacrifice ... It will turn out to be a fruitful death. An event of human

history will make its mark in the annals of the people. The legal murder of Ravachol will open up an Era. And you artists who daub on canvas your mystical dreams, your fluent brush is offered the great subject of your work. If you have understood your epoch, if you have recognized and kissed the threshold of the future, you must draw the life of the Saint, and his passing, in a pious triptych. For there will be a time when in the temples of the Real Brotherhood, you will fit your stained glass in the most beautiful place, so that the light from the Sun passes in the halo of the martyr on a planet that is free from property![37]

Victor Barrucand, similarly, worshipped Ravachol as a kind of 'violent Christ'. 'Ravachol', he wrote, 'will perhaps appear one day as a sort of violent Christ, such as the time and milieu he passed through could produce'. Both Ravachol and Christ

> wanted to destroy wealth and power, not to seize them. One preached gentleness, spirit of sacrifice and renunciation ... the other preached, through example, revolt against abusive authority, individual initiative against the cowardice of the masses, the claim of the poor to earthly happiness ... they taught the world that the ideas of fatherland and society, any more than those of worship and law, cannot prevail against the right of man to be happy – in this world, said Ravachol; in Heaven, said Jesus.[38]

We are a long way here from the joyful, affirmative humour of the Montmartre cabarets. Yet Barrucand's article is dominated by a striking image of Ravachol's severed head gazing from the foot of the guillotine at the crowd of spectators. Frozen upon the head is a laugh: a cold, bitter laugh, with insults and derision on its lips. Barrucand reflects on this laughter at some length:

> Oh! This laughter before the sinister machine – Homeric derision reflected in the silence of the summer morning when all life wanted to smile! – It gives a funereal shudder; and the social whore at the foot of the scaffold, attacked by the sarcasm of a criminal who has no concern with politeness but brings into action a surprising energy, is forever withered, as if deservedly spat in the face for all its infamy and catastrophic mediocrity. His jeering blasphemy unalterably frozen by the rictus of death, the head of the revolt, beautiful and purified, remains, with I don't know what legendary authority.[39]

This evocation of new forms of secular, anarchistic religiosity partly subverted the Positivist philosophy of writers like Comte and Littré, who attempted to find a scientific foundation for social and political authority in a 'positive' stage of history that would supersede the earlier theological and metaphysical stages. Positive philosophy, Comte wrote, aimed to realize organic qualities of precision, certainty, usefulness, reality and relativity. A positivist study of real facts could 'direct the spiritual reorganization of the civilized world' by redirecting the worship of God towards the worship of humanity itself.[40] Anarchist thought was influenced by the spirit of positivism's attempt to find a scientific foundation for social and political authority (as evidenced by the scholarly work of academic geographers such as Kropotkin and Reclus), but held that realising the organic qualities of positive knowledge would reveal an autonomous, anti-authoritarian society as being the most scientifically rational way of organising the world. In the sanctification of Ravachol, we clearly see the influence of positivism's evocation of a new religion of humanity, but turning this upside down so that it was the greatest *opponent* of existing society who was to be mankind's secular Redeemer.

In performing this ironic beatification of revolutionary violence, it seems to me that anarchists performed a new kind of reactivation of the Cynic King of derision. Barrucand's description of Ravachol dwells on the lifeless laughter of Ravachol's severed head, its jeering challenge to the cheering spectators of state murder. In his death, Ravachol poses a powerful challenge to the inauthentic and degenerate nature of modern societies. His laughter in the face of death reveals, not only his bravery, but the authority of his 'parrhesia', his speaking truth to power. Ravachol's laughter is a laugh of derision and hatred. It reverses hierarchies, proclaiming himself as the true king. Yet it is a form of parrhesia that has broken down and descended into violence; the mutual relations of care between the speaker and the addressee have collapsed. Ravachol's severed head marks the failure of the attempts to find new ways of speaking militant truths with authority, and the falling back to a politics of mutual antipathy and violence.

In violence, anarchist laughter remains present, but it is a laughter that draws from irony, not humour. This is laughter as a sheer vital force, one that is made more powerful through the authenticity and certainty of death. This is certainly not a form of laughter that descends to a play of surfaces; rather, it is a form of laughter that elevates the subject and the exalted principles he stands for. In the Deleuzian framework discussed earlier, Ravachol's laughter can be viewed as an ironic form of cruelty that refers back to transcendent morality and subjectivity. The hagiography of Ravachol attempts to create a new sacred foundation for a new, anarchist religion of humanity. In doing so, Ravachol's laughter is closer to the modern cynicism of irony than the affirmative materialism of humour. It attempts to create new foundations and new grounds. In doing so, its violence becomes veiled and implicit even at the moment it is most visible, since it is now cloaked in a new transcendent morality.

This turn towards transcendent foundations and grounds is discernible elsewhere in the anarchist culture of the 1890s, in particular anarchists' nostalgia for 'primitive' societies. One only moderately tongue-in-cheek pamphlet that was collected in the police archives, for example, imagined a 'primitive' man, magically transported to the present day, in conversation with four men of modernity: a mine worker, a factory labourer, an agricultural worker, and an office clerk (see Figure 4.2).[41] As the group stand in front of symbols of French modernity such as the Eiffel Tower, the primitive man, stupefied at

Figure 4.2 Anonymous, Front page of the journal *L'Etat naturel et la part du prolétaire dans la civilisation*, 1894, Paris

seeing their unhealthy physique, asks them a series of questions: 'Why these black marks all over your white face? Why are you so thin and enfeebled? Why do you have no hair or teeth? Why are you so broken and tired?' The pamphlet goes on to lyrically describe the happy life of primitive man, with his easy life of hunting, dancing and abundant sex, not to mention his average life expectancy of 120 years. 'And now, Sociologists', the article concludes, 'compare this with the life of the Proletarian in civilization!'[42] Pre-modern man, the article emphasises with gentle humour, enjoyed a vigorous vitality that has withered in the modern age.

This primitivism also helps explain the theory of communication on which propaganda by the deed – a form of 'propaganda' with no concrete representational content – implicitly relied. There were close ties between anarchists and the Symbolist avant-garde, which was based on a fascination with exploring the hidden, mysterious, primeval energies lurking beneath the everyday order of things.[43] The poetry of Mallarmé, for example, unravelled the structures of linguistic representation, abandoning conventional narrative and description whilst seeking new forms of meaning in the flights of association and imagination beyond the word's 'practical' use of signification.[44] Mallarmé paid close attention to the embodied aspects of language, evoking new meanings through novel use of textures of sound, rhythm and symbol. His work was taken up by a school of followers who attempted to create forms of symbolic poetry which, as Jean Moréas put it, 'clothe the Idea in a sensible form which, nonetheless, will not be an end in itself, but which, while serving to express the Idea, remains the subject'.[45]

This impulse to 'dress the Idea in sensible form' is clearly discernible in the anarchist political culture of propaganda by the deed. Fin-de-siècle anarchists emphasised the ways in which violence made ideas more vital, more living.[46] 'The idea', wrote Georges Brousse, 'will not appear on paper, nor in a newspaper, nor in a painting; it will not be sculpted in marble, nor carved in stone, nor cast in bronze: it will walk, alive, in flesh and bone, before the people. The people will hail it as it passes.'[47]

Through violence, thought was not only to awaken (as Kroptokin had written), but to come alive. This celebration of excessive energy in all its forms achieved expression in an anarchist song, widely sung and danced to in Montmartre cabarets, called *Lady Dynamite:*

> Our fathers once danced
> To the sound of the cannons of the past!
> Now this tragic dance
> Requires stronger music.
> Let's dynamite, let's dynamite!
> Lady Dynamite, let's dance fast!
> Let's dance and sing!
> Lady Dynamite, let's dance fast!
> Let's dance and sing, and dynamite!

Humour's vitality has given way here to a different kind of vitality, an excessive energy of dance and violence. In fin-de-siecle medical science, laughter and dancing were both widely regarded as expressions of animalistic vital urges (and associated with creative maladies such as hysteria), and in this context became associated with an anarchic liberation of natural energy and vitality.[48] Anarchist culture now added dynamite to laughter and dance as means of liberating what Charles Baudelaire had referred to in 'The Painter of Modern Life' as life's 'luminous explosion in space'.[49]

In the doctrine of propaganda by the deed, we can see, anarchism lost its orientation towards a positive affirmation of humour as (in Deleuze's vocabulary) a form of 'cruelty' whose material force ruptures vertical structures of meanings, authority, law and subjectivity. Instead, it fell back on a more cynical attitude in which humour's ambivalence, ambiguity and play of surfaces gave way to the certainties of an ironic, vertical structure of signification that referred back to authentic origins and redemptive acts of mystical violence.

Conclusion

Fin-de-siècle anarchist culture, I have shown, demonstrated varying attitudes towards humour, laughter, cruelty and violence. As frustration grew with the capacity of cultural texts and performances, including the humorous texts and spectacles of the anarchist cabarets of Montmartre, to prompt revolutionary change, the dominant tone of anarchist culture shifted. As the tensions inherent within the attempt to live out alternative, utopian forms of life in Montmartre became clearer and more problematic (due not least to the surprising commercial success of many Montmartre cabarets that celebrated bohemian or down-and-out culture), the doctrine of propaganda by the deed offered a more immediate and far less ambiguous alternative. Humour, as a form of 'cruelty', started to give way to forms of physical violence that were immediately interpreted through logics of transcendence, redemption, and authenticity.

Whilst anarchist humour did little to convince the bourgeois audience it attacked, it did set an agenda for a continuing affinity between humour and radical politics in coming decades. The Surrealist artist André Breton's *Anthology of Black Humor*, for example, is filled with anarchist writers, whilst recent autonomist political practice has seen a resurgence of humour-based political action. Such humour has most potential, however, when it remains as a practice of cruelty – generating new events of thought and feeling – rather than as a practice of transcendent morality, communicating pre-digested truths, or as a practice of violence.

Notes

1 Bihl, '"L'Armée du chahut": les deux Vachalcades de 1896 et 1897'; Weisberg, *Montmartre and the Making of Mass Culture*; Cate, 'The Spirit of Montmartre', in

Cate and Shaw (eds), *The Spirit of Montmartre: Cabarets, Humor and the Avant-Garde, 1875–1905*.

2 Sharpe, Hynes and Fagan, 'Beat Me, Whip Me, Spank Me, Just Make It Right Again: Beyond the Didactic Masochism of Global Resistance'.

3 Routledge, 'Sensuous Solidarities: Emotion, Politics and Performance in the Clandestine Insurgent Rebel Clown Army', 428–429.

4 Kanngieser, *Experimental Politics and the Making of Worlds*.

5 See also Genosko, *Félix Guattari: An Aberrant Introduction*.

6 See Critchley, *On Humour* and Critchley, *Infinitely Demanding: Ethics of Commitment, Politics of Resistance*.

7 Billig, *Laughter and Ridicule: Towards a Social Critique of Humour*; Kuipers, *Good Humor, Bad Taste: A Sociology of the Joke*; Lockyer and Pickering (eds), *Beyond a Joke: The Limits of Humour*; Powell and Paton (eds), *Humour in Society: Resistance and Control*; Weaver, *The Rhetoric of Racist Humour: US, UK and Global Race Joking*.

8 Billig, 'Humour and Hatred: The Racist Jokes of the Ku Klux Klan'; Billig, 'Comic Racism and Violence', in Lockyer and Pickering (eds) *Beyond a Joke*.

9 Ridanpää, 'Geopolitics of Humour: The Muhammed Cartoon Crisis and the Kaltio Comic Strip Episode in Finland'; Weaver, 'Liquid Racism and the Danish Prophet Muhammad Cartoons'.

10 Colebrook, *Irony*, 130.

11 Merdauculatives guenilles,
Guenipes aux regards malsains,
Gaupes, gouges, vadrouilles, filles
Dont s'honorent nos traversins,
Souffrez que notre main caresse
La gélatine de vos seins,
Tirependières en détresse,
Sous les bosquets des Assassins.

Potaches, rapins, joyeux drilles,
Larbins, collignons, petits saints
Portant casaques et mandilles,
Capitans, bretteurs, spadassins
Qu'arde la soif, cette traitresse,
Accourez ici par essaims
Communier à la dive messe
Au maître-autel des Assassins.

Poètes rongeurs de croustilles,
Mieux dentés que des marcassins,
Là d'alléchantes béatilles
Remplissent d'argileux bassins;
Pipeurs de bémols en liesse
Qui chatouillez les clavecins,
Venez cueillir des fleurs d'ivresse
En bénissant les Assassins.

Citadins amants de courtilles
Venez ici, loin des roussins,
Livrer votre joie aux charmilles,
Sous les regards des Assassins.

12 Harvey, 'Monument and Myth'; Jonas, 'Sacred Tourism and Secular Pilgrimage: Montmartre and the Basilica of Sacré Coeur', in Weisberg (ed.), Montmartre and the Making of Mass Culture.

84 *Julian Brigstocke*

13 Sonn, *Anarchism and Cultural Politics in Fin-de-siècle France.*
14 Brigstocke, *The Life of the City*: Space, Humour, and the Experience of Truth in Fin-de-siècle Montmartre.
15 Sonn, *Anarchism and Cultural Politics in Fin-de-siècle France.*
16 Brigstocke, 'Defiant Laughter: Humour and the Aesthetics of Place in Late 19th Century Montmartre'.
17 See the excellent discussion in Jones, *Sad Clowns and Pale Pierrots*: Literature and the Popular Comic Arts in 19th-Century France.
18 Cited in Jones, *Sad Clowns and Pale Pierrots*, 194.
19 Salis, for the 'Poster Montmartre municipal election'.

> Qu'est-ce-que Montmartre? – Rien ! Que doit-il être ? – Tout ! Le jour est enfin venu où Montmartre peut et doit revendiquer ses droits d'autonomie contre le restant de Paris. En effet, dans sa fréquentation avec ce qu'on est convenu d'appeler la capitale, Montmartre n'a rien à gagner que des charges de ses humiliations. Montmartre est assez riche de finances, d'art et d'esprit pour vivre sa vie propre … Montmartre mérite d'être mieux qu'un arrondissement. Il doit être une cité libre et fière.

20 Lehardy, 'Montmartre'.

> Montmartre est isolé, parce qu'il se suffit à lui seul. Ce centre est absolument autonome. Dans une petite ville située à une grande distance de Montmartre et que les voyageurs nomment Paris, une académie locale discute, dit-on, les conditions de l'autonomie municipale. Il y a longtemps que cette question est résolue à Montmartre.

21 Wilson, *Paris and the Commune, 1871–78.*
22 Cited in Sonn, *Anarchism and Cultural Politics in Fin-de-siècle France*, 130.
23 Un jour louche tombait du plafond et quelques lanternes aux vitres sales étaient accrochées aux piliers … dès que la porte s'ouvrait, chacun était reçu par des bordés d'injures et 'en prenait pour son grade' comme on dit aujourd'hui. C'était une vraie fontaine de purin…. Les garçons avaient tous des airs de bandits, une barbe de trois jours. Ils savaient marcher dans les entraves avec leurs gros sabots. Mais parfois, ils se penchaient sur les épaules des belles dames en respirant leurs parfums pour donner à ces colombes le frisson de la guillotine.
24 Lisbonne, 'La Taverne du Bagne'.

> Entre Paris et Montmartre, … l'ex-forçat Maxime Lisbonne, vient de ressusciter et de résumer le Bagne. C'est une hardiesse et une curiosité unique dans l'histoire des fantaisies qui ont rendu fameuse la Butte chère aux Parisiens … Le personnel, attaché au service de la Taverne, a été choisi parmi des anciens Fonctionnaires, Négociants, Industriels, Financiers, Propriétaires, Prêtres, Frères, Ignorantins, qui, ayant subi leur peine qu'à vivre honnêtement … Vous tous, que êtes entrés au Bagne – et qui, pourtant, en êtes sortis – merci pout la constante bienveillance avec laquelle vous avez traité les forçats … Vous avez aidé au relèvement des déchus, à la moralisation des dévoyés.

25 Wilson, *Bohemians: The Glorious Outcasts.*
26 Grand-Carteret, *Raphael et Gambrinus.*
27 Brigstocke, 'Artistic Parrhesia and the Genealogy of Ethics in Foucault and Benjamin'.
28 Foucault, *The Courage of Truth*: The Government of Self and Others, 284.
29 Foucault, *The Courage of the Truth*, 278.

30 Sloterdijk, *Critique of Cynical Reason*, 101.
31 Sloterdijk, *Critique of Cynical Reason*, 103.
32 Goudeau, *Dix ans de bohème*.

> [Salis] dut revêtir un costume en or, des étoffes inouïes, se munir d'un sceptre. Après avoir reçu les hommages des peuples, il s'en alla prendre possession du Moulin de la Galette. Il s'y rendit, cachant ses vêtements royaux sous un ulster, accompagné par des peintres et des poètes armés de hallebarde, qui, tout le long de la butte, à l'ahurissement des populations, criaient: Vive le roi!

33 Cole, 'Dynamite Violence and Literary Culture', 301.
34 Sergey Nechayev, *The Revolutionary Catechism*; available at http://www.marxists. org/subject/anarchism/nechayev/catechism.htm (accessed 9 September 2016).
35 Préposiet, *Histoire de l'anarchisme*.
36 Anon. "News of the Week", The Spectator (16 July 1892), pp. 1–3.
37 Adam, 'Eloge de Ravachol', 27–30.

> Ravachol reste bien le propagateur de la grande idée des religions anciennes qui préconisèrent la recherche de la mort individuelle pour le Bien du monde; l'abnégation de soi, de sa vie et de sa renommée pour l'exaltation des pauvres, des humbles. Il est définitivement le Rénovateur du Sacrifice Essential ... Une mort féconde va s'accomplir. Un évènement de l'histoire humaine va se marquer aux annales des peuples. Le meurtre légal de Ravachol ouvrira une Ère. Et vous artistes qui d'un pinceau disert contez sur la toile vos rêves mystiques, voilà offert le grand sujet de l'œuvre. Si vous avez compris votre époque, si vous avez reconnu et baisé le seuil de l'Avenir, il vous appartient de tracer en un pieux triptyque la vie du Saint, et son trépas. Car un temps sera où dans les temples de la Fraternité Réelle, on emboîtera votre vitrail à la place la plus belle, afin que la lumière du soleil pass[e] dans l'auréole du martyr sur la planète libre de propriété!

38 Barrucand, 'Le Rire de Ravachol'.

> Ils voulaient tous les deux, ces démolisseurs du temple, anéantir la richesse et le pouvoir, et non pour s'en emparer. L'un prêcha la douceur, l'esprit de sacrifice et de renoncement, l'affranchissement des entraves politiques par le dédain, en vue de la conquête du royaume céleste; l'autre prêcha par l'exemple la révolte contre l'autorité abusive, l'initiative individuelle contre la lâcheté des masses, la revendication des pauvres au bonheur de la terre. Affranchis de l'égoïsme étroit, ils prirent une plus haute conscience d'eux-mêmes dans l'humanité; partis du principe d'amour, malgré d'apparentes contradictions, ils marchèrent à leur but; avec une volonté héroïque, ils enseignèrent au monde que les idées de patrie et de société, non plus que celles de culte et de Loi ne sauraient prévaloir contre le droit qu'à l'homme d'être heureux, dans ce monde a dit Ravachol, dans le ciel a dit Jésus.

39 Barrucand, 'Le Rire de Ravachol'.

> Oh! ce rire devant la sinistre machine – homérique dérision répercutée dans le silence de l'estival matin où toute vie voulait sourire! – il donne le frisson funèbre; et la catin sociale, au pied de l'échafaud, atteinte par le sarcasme et le défi d'un criminel peu soucieux de politesse mais qui apporte dans l'action une énergie si surprenante, est à jamais flétrie, comme si toutes ses infamies et son irrémédiable médiocrité lui étaient jetées à la face dans ce crachat qu'elle a

mérité. Son gouailleur blasphème immuablement figé par le rictus de la mort, la tête du révolté, belle et purifiée, demeure, avec je ne sais quelle autorité légendaire.

40 Comte, *A General View of Positivism*, 48.
41 Anon., 'L'état naturel et la part du prolétaire dans la civilisation'.
42 'Et maintenant, Sociologues, comparez à cela, l'existence du Prolétaire dans la Civilisation!'
43 On anarchism and Symbolism, see Hyman, 'Theatrical Terror: Attentats and Symbolist Spectacle'; Aubery, 'The Anarchism of the Literati of the Symbolist Period'; Shryock, 'Becoming Political: Symbolist Literature and the Third Republic'; Williams, 'Signs of Anarchy: Aesthetics, Politics, and the Symbolist Critic at the Mercure de France, 1890–1895'.
44 For a discussion of the echo of Rousseau in Symbolists' approach to language, see the first chapter of Candida Smith, *Mallarmé's Children*: Symbolism and the Renewal of Experience.
45 Moréas, 'Le Symbolisme'. '[L]a poésie symbolique cherche à vêtir l'Idée d'une forme sensible qui, néanmoins, ne serait pas son but à elle-même, mais qui, tout en servant à exprimer l'Idée, demeurerait sujette'.
46 Sorel, *Reflections on Violence*.
47 Brousse, cited in Hyman, 'Theatrical Terror'.
48 Gordon, 'From Charcot to Charlot', 529.
49 Baudelaire, 'The Painter of Modern Life'.

PART II

Early anarchist geographies and their places

5 The thought of Élisée Reclus as a source of inspiration for degrowth ethos

Francisco Toro

The theory of degrowth has emerged in the last decade as a bottom-up alternative to the dominant, hierarchical and centralised discourse of sustainable development. One of its main mottos is 'decolonising our imaginary'[1] from the growth myth and material accumulation of capitalism and consumerism. Main initiatives have occurred in Mediterranean and Southern regions of Europe. The foundations of contemporary degrowth are the result of the convergence of two strands of radical thought.[2] On one hand, the 'ecological critique', based on the bioeconomics of Nicolas Georgescu-Roegen (1906–1994), the first author who suggested the analysis of economic systems according to the law of entropy; and early twentieth century thinkers such as P. Geddes (1854–1932), S. Podolinsky (1850–1891) and F. Soddy (1877–1856) who advanced the idea of bioeconomics. On the other hand, the 'culturalist critique', which encompasses the radical critique of modernity and the techno-industrial society, whose main intellectual milestones are A. Gorz (1923–2007) (he first used the term 'degrowth'), Ivan Illich (1926–2002), J. Ellul (1912–1994), B. Charbonneau (1910–1996), in which we may include the post-development approaches.[3] Is there any room for Élisée Reclus, knowing his ties and connections with a radical and critical side of progress and environmentalism?

The reflection on environmental implications of human activity on Earth, the critique of capitalist rationality and injustice, the idealisation of self-managed communities, the idea of 'good life', and the advocacy of human welfare that reinforces qualitative dimensions, are immanent in some of Reclus' essays, being basic precepts of what degrowth philosophy should be. This chapter proposes a review of some of these insights and their potential inspiration for degrowth movements and traces a line of connection with a radical and anarchist side of environmentalist and more transformative geographic thought.

For showing and analysing the presence of a 'degrowth' discourse in Reclus' thought I will use a structure according to the synthesis of the main degrowth topics related to deep and long philosophical themes in the Western thought, but also (and, even with a greater presence) in other cultural influences. These five 'sources' of degrowth are[4]: i) ecology, ii) critiques of development and praise for anti-utilitarianism, iii) meaning of life and well-being, iv) bioeconomics,

v) democracy, and vi) justice. Finally, I will conclude with a summary of his main contributions, justifying him as a determinant figure who ought to inspire the contemporary strands of degrowth.

Is Reclus a missing link in the course of degrowth theory?

Though degrowth has appeared as a need of our times, specifically for reacting to the oxymoron and non-transformative paradigm of sustainable development, the reflection on the idealisation of balanced societies with nature is not a new theme within western thought. Consequently, it may extend the research on the historical roots of degrowth by looking at thinkers and scholars who pre-date radical approaches of the twentieth century.

In this regard, A. Sippel has found some similarities between contemporary degrowth partisans and the eighteenth- and early nineteenth-century utopists.[5] Notable figures such as William Godwin and Jean-Jacques Rousseau advocated frugality over luxury, a voluntary reduction of material goods in daily life, the importance of education for transforming and *decolonising* permanently unsatisfied minds, and the idealisation of small communities and 'eco-villages' as the symbol of a balanced encounter between humans and nature. These insights would completely match contemporary degrowth purposes. In this regard, A. Sippel concludes that

> today's *décroissants* are closer to Godwin and Rousseau's socialism than to later nineteenth-century social thinkers, as the former advocated a refusal of luxury as a means of establishing equality rather than the expectation that further development should be shared and thus push up all living conditions.[6]

It is well known that Godwin and, to a lesser extent, Rousseau, influenced the thought of anarchist geographer Élisée Reclus. For instance, Reclus shared with Godwin the idea that 'community and solidarity can never be separated from liberty and individuality'.[7] Rousseau's conception of 'noble savage' probably suggested to Reclus and other nineteenth-century anarchists the advocacy of a harmonic integration of humans into nature, as long as human freedom is guaranteed.[8] Did they also inspire Reclus in terms of a degrowth *philosophy*? Is degrowth (or something analogous) a theme cultivated by Reclus and other early anarchist geographers?

Several authors agree in defining Reclus as a proto-environmentalist, being acknowledged as a source of inspiration for environmental ethics, vegetarianism and radical ecological social thought[9] akin to the ecologist approach of degrowth thinkers. B. Giblin regards him as an *avant l'heure* environmentalist. Along with other early anarchist geographers, such as Kropotkin and Metchnikoff, Reclus showed 'that Earth is a living planet where human actions have both negative and positive effects and they depend upon the political and economic system present'.[10] Thereby, their critiques of environmental degradation are organised and structured, stressing the importance of the political and economic order in the use and management of nature.

Though the global environmental system has significantly changed in the last century, Reclus' reflections on the human impact on the environment seem still relevant today.[11] His concepts, insights, reflections and critiques were signs of a radical environmentalism, which was based on a complex way of looking at reality, challenging the hegemonic theories around science, progress, politics and economics of that time. Reclus insisted on the need to care for the planet, seeking harmony between human needs and the physical possibilities of nature to meet them, whereby the association between social injustices and environmental degradation could be addressed through the pedagogical values of a geographical approach that emphasised integrative, holistic and emancipating knowledge.[12] The coalescence of environmental and social approaches is thus embodied in Reclus. As it may seem, there are quite a few points in common with the partisans of degrowth, who are involved in a permanent struggle against the hegemonic theories, discourses and policies of the development and growth axiom.

Moreover, the main exponent of degrowth, Serge Latouche, refers to Reclus as a thinker who followed in certain ways the degrowth philosophy in his thought: 'degrowth has affinity with the first inspiration of socialism, as it was pursued by independent thinkers such as Élisée Reclus and Paul Lafargue'.[13] All of the previous arguments would fairly justify the exploration of Reclus' thought as a potential source of inspiration for contemporary degrowth.

Unfortunately early anarchist geographers were not well enough acknowledged in the foreground of ecological thought during the twentieth and twenty-first century, even in anarchist thought. As Clark and Martin argue, 'ecological thinking remained an undercurrent of anarchist and utopian thought and practice... [yet] it did not become a central theme in anarchist and utopian theoretical discussion until the ideas of Paul Goodman and Murray Bookchin began to have a noticeable influence in the late 1960s'.[14]

Yet, Reclus has been of less interest than Kropotkin who is considered to have exercised more influence on later radical thought, even in an environmentalist dimension.[15] Certainly, Reclus' works are inspired by philosophic grounds which clearly recall basic arguments of a biocentric and radical contemporary environmentalism. In overall terms, the roots of his thought are immersed within a green conception of human welfare, and express a more balanced treatment of the tension between human progress and environmental protection, opting very often for the latter.[16] Reclus' thought is characterised by a non-utilitarian understanding of nature, although he did not deny the substantial potential benefits of the enhancement of physical attributes and natural resources thanks to great advances in technology and science.

According to Giblin, the environmental sensibility of Reclus had no immediate posterity,[17] and began to be rediscovered in the 1960s among anarchist thinkers and scholars,[18] remaining little known for most of the critical community of the contemporary environmentalist movement.

Curiously, this re-emergence of Reclus among anarchists was not always lauded.[19]

Degrowth ethos in the thought of Reclus

Ecology

Ecology provides degrowth with determinant foundations that drive to surpass the reductionist understanding of nature according to neoliberal economics. An ecological perspective would conceive of natural resources and services as compounds of complex and non-linear natural systems. Accordingly, they play an important role in the sustainability and maintenance of ecosystems, habitats and natural cycles. First and foremost, intrinsic values are acknowledged in nature and its attributes, beyond monetary parameters and utilitarian demands. Based on this scientific precept, the ecological approach would warn of the incompatibility between large-scale industrial production and consumption systems and the physical viability of ecosystems. In this regard, degrowth would be 'a possible path to preserve ecosystems by the reduction of human pressure over ecosystems and nature, and a challenge to the idea that decoupling of ecological impacts from economic growth is possible'.[20]

Precisely, the first essays of Reclus were written in times of the emergence of Ecology as a science aiming at discovering the interactions between species and their environment. Yet, this early Ecology had hardly anything to do with the social and political implications that the ecological paradigm has reached for degrowth. Advances in natural sciences throughout the twentieth century, such as chaos theory, complexity in non-linear systems, new thermodynamics, will open new perspectives in Ecology, overcoming mechanist understandings of nature. Such scientific background is crucial in the environmentalist discourse of degrowth theory. Thus, degrowth complements the revolutionary shift that the ecological paradigm has produced in our understanding of *Gaia*. It encompasses a wide range of theoretical approaches and political movements whose *raison d'être* is Ecology: Deep Ecology, Social Ecology and Ecofeminism, all of which, according to Graham Purchase, 'are... inherent in anarchist philosophy'.[21] Nevertheless, early anarchist geographers like Reclus were reluctant to adopt both the term 'Ecology' and its content,[22] insofar as it was identified with social Darwinism, made notorious by one of its founders, Ernst Haeckel.

The theorisation of human–nature relationships in Reclus was thus driven separately from: i) the genealogy and development of Ecology; ii) its most notable authors; iii) its ideological connotations. Instead, the leading light of Reclus' thought would be 'mesology', to which he did not refer explicitly until his last and most famous work *L'Homme et la Terre* (1905–1908). Mesology is a direct influence of his readings of Louis-Adolphe Bertillon, who, in turn, inherited the idea from August Comte[23] who studied the interactions between humans and their physical and social-cultural *milieux* in an interdisciplinary way. Yet, Reclus never quoted Bertillon, but others who moved in the same vein, such as H. Drummond (1851–1897), R. von Ihering (1878–1892) and G. de Greef. The key concept, *milieu*, would fit perfectly in the construction of what Reclus called 'social geography'.[24] Unlike his contemporary Ecology, the idea of environment ('milieu') used by Reclus is plural, diverse and historical; a sign of the multiple varieties of how humans have been adapting to the physical features of the environment. Within the apparent neutrality in which science works, the idea of environment is another proof of how intentionally ideological science and scientists can be. Whilst the dominant vision of nineteenth-century Ecology was to emphasise an inevitable hierarchy in the structure and functions of wildlife and nature, mirroring and justifying the imbalances among human groups, Reclus rejected this organising principle and substituted it with 'affinity',[25] searching for harmony and complementarity as the main tenet, to inspire more equal and fair societies. In short, both approaches and denominations create different discourses from the same reality.

In this regard, the mesology of Reclus draws on the comprehension of complexity in the reciprocal and emergent relationships between human and nature. Such understanding would be determinant in three basic precepts of degrowth ethos: i) the carrying capacity of human activities; ii) the inevitable omnipresence of physical laws in human works; iii) and their material dependence on natural resources and services. Yet, his historical context was not favourable for such kinds of arguments, due to the incredible technological gap gained by dominant powers in the nineteenth century. This dissident attitude is equivalent to that of degrowth partisans within the era of globalisation and neoliberalism. As Clark and Martin argue: 'Reclus lived in an age in which social analysis tended toward either an idealism in which material determinants were ignored or a materialism in which economic and technological determinants were attributed almost exclusive importance'.[26]

Social-Darwinist strands of Ecology have been used in justifying inequalities among peoples and proclaiming the racial superiority of selected cultures over others according to natural causalities and biological principles. In fact, E. Haeckel became one of the main German ideologists of racism, nationalism and imperialism.[27] Unlike social-Darwinists, Reclus would use these causalities for finding intimate relationships between a human group and its surroundings. Though inserted in his time, Reclus sought to emphasise the close connection between conditions such as climate, topography or the aesthetic qualities of the landscape and the physiological and behavioural characteristics of its dwellers. Thereby, he anticipated modern environmentalism by showing that nature plays a decisive influence not solely on the *modus vivendi*, but also in cultural and religious manifestations. This is a plausible way to stress the need for *caring* for a humanised and historical nature according to each human group and territorial context, and into which the inhabitants have virtually merged. According to Pelletier, in Reclus' work, 'a supposed determinism is immediately compensated by variety'.[28]

For Reclus, the coevolution between Man and Nature and the search for a reciprocal harmony is one of the basic laws of a 'social geography'.[29] Not surprisingly, he warned of the capacity of humankind for degrading the physical environment on a global scale. Such a view was determinant of his review of G. P. Marsh's work, *Man and Nature* (1864), entitled *L'homme et la nature: de l'action humaine sur la géographie physique* (1864). He claimed that 'action of man may embellish the earth, but it may also disfigure it; according to the manner and social condition of any nation, it contributes either to the degradation or glorification of nature'.[30] The need to transform physical features into landscapes, buildings and artificial environment is an immanent feature of humankind, common to all groups and cultures, surpassing the intensity of any other animal to transform the Earth: 'Man's actions, on the contrary, have greatly changed the appearance of the surface of the earth'.[31]

In the latter case, Reclus used the term 'man' generically, whilst, it has been said above, his conception of environment is diverse and plural, historical and cultural. Is it contradictory or intentional? One might think that his intention was not to show environmental problems as anecdotal or limited to specific territorial contexts, but rather to analyse them as something embedded in the technical progress, intellectual background, behaviours and moral principles of humans as a whole, i.e., the human race. In doing so, he anticipated that the global scope of anthropic impacts would reach the most remote parts of the Earth: 'The day is approaching when there will remain no region on any continent that has not been visited by a civilised pioneer, and sooner or later, the effects of human labour will extend to every point of the surface of the earth'.[32]

One of the biggest and most confusing dilemmas for degrowth thinking is the matter of population growth. If degrowth demands a cessation of any material human progress in order to avoid ecological collapse, would that also require a severe decrease in world population? Following pure ecological laws, it is easy to foresee the 'tragedy' of ecosystems when an overpopulation of individuals fights desperately for access to scarce food and land. As a sort of strictly biological comparison of humans with others animals, this scheme inspired the work of the ecologist Paul R. Ehrlich in the 1960s and 1970s, which is reminiscent of Malthusian arguments.[33] |Moreover, a terrain of ambiguity might be projected in the degrowth movement, as what concerns population growth and its pressure over the carrying capacity of the Planet has rarely been discussed in degrowth theory.[34] Yet, degrowth itself, according to left and radical positions, reproduces discourses opposed to population growth as a determinant factor of environmental crisis, instead putting more emphasis on social inequality in consumption per capita.[35] Though neo-Malthusian theory has a strong echo amongst *degrowthers*, it is more a political option to struggle against the capitalist exploitation of female bodies to produce soldiers and cheap labour,[36] than a reaction to an ecological problem of over-population. Neo-Malthusian arguments within Iberian anarchism advocate a conscious procreation and free maternity, promoting feminism and a new sexual ethics.[37] This side of degrowth is clearly inherited from anarcho-feminists such as Emma Goldman. But, equally, this approach may find some source and a similar spirit within early anarchist thinkers, in particular, Reclus.

Reclus considered population growth a determinant factor both in the rise and the fall of societies, and somehow may affect their future viability: 'growth in numbers has been, without doubt, an element contributing to civilization, it has not been the principal one, and in certain cases it can be an obstacle to the development of true progress in personal and collective well-being, as well as to mutual good will'.[38] Yet, he was very critical about Malthusian theories, as they justified deep imbalances within the modern society. According to Reclus, starvation, insalubrity and insecure conditions in the workplace were not misfortunes, but they responded to an economic and productive model which generates and feeds these social imbalances among different social classes. Thus, Reclus did not hesitate to question the scientific foundations of liberalism, a doctrine which drew on Malthusian arguments considering the death from hunger of poor people as unavoidable.

Reclus trusted in the capacity of technology for increasing food production without aggressively harming the environment, recalling the current approach of eco-efficiency. But from his point of view, this increase of global production had to be linked to non-capitalist agriculture and enable self-consumption among the community of producers. In effect, there would be enough natural resources for ensuring the basic demands of the population, which is closer to the idea of self-sufficiency, a basic principle in degrowth philosophy:

> The land is large enough to keep us on its breast, and rich enough to afford us living comfortably. It may give enough crops for feeding every-one, makes grow enough fibre plants for dressing everyone, and contains enough stones and clay so that everyone may have homes. That is the economic fact in all its simplicity.[39]

Critiques of development and praise for anti-utilitarianism

For degrowth theory, the term *development* colonises our imaginary[40] and makes it difficult to think other ways of enhancing the conditions of life outside growth. The equivalent of this term, according to cultural inheritance of the Enlightenment is the idea of *progress*, which is central in Reclus' thought. Since the second half of the twentieth century, the idea of development replaced the meaning that progress has in Modernity, reducing it to its economic dimension. Then, many attempts were made to correct the problem of its monetary bias by qualifying development with countless adjectives such as 'local', 'human', 'social', 'sustainable', etc.[41] At present, progress remains as the core of development and its use is devoted to emphasise the scientific and technological conquest of so-called developed countries.[42]

Unlike degrowth's partisans, Reclus did not refer specifically to the model of economic growth, as this axiom began to be challenged by critical discourse in 1950s to 1960s, after its institutionalisation and legitimation as an imperative for the development of both rich and poor countries. But he wrote about qualities, attitudes and values of the human condition that negatively affect the harmonic relation between societies and their environs. In this regard, his reflections fit perfectly with degrowth purposes, insofar as he made a critique of intrinsic characteristics of capitalism, such as dissatisfaction, egoism or luxury, though he rarely explicitly alluded to this economic system. Reclus foresaw the rise of a world economic system impregnated by a global culture based on economistic values, or what degrowth partisans refers to as the 'colonisation of imaginary'. Also, he witnessed the beginning of a global extension of capitalistic ideology all over countries, including the old European colonies: 'those countries of Asia that have developed in the direction of the ideal world of economics, and in all other parts of the world that are carried along by the example of Europe and its all-powerful will'.[43]

Regarding the utilitarian attitudes of the human towards nature, Reclus' insights are intended to be moralistic and pedagogical. He blamed superfluous or overexploiting human attitudes against nature: 'it can be said that man, jealous of nature, tries to belittle the products of the soil and does not allow them to surpass his level'.[44] In a certain way, he is reflecting pedagogically on the notion of limits, insofar as the demands are not enough smothered with the possibilities that nature gives to humans. As Clark and Martin assert,

> his position on this issue is very similar to that of many contemporary social ecologists who concur with Reclus that human society has throughout history substituted one form of social hierarchy for another and has increasingly adopted an exploitative and destructive standpoint toward the natural world.[45]

Moreover, Reclus rejected any kind of attempt at appropriating and exploiting the qualities of nature, such as wildlife or beautiful landscapes, considering that these qualities are free and common to all humans. He reasserted thus his total opposition to the commodification of nature, above all if it is not justified in order to solve basic and urgent needs: 'At the seashore, many of the most picturesque cliffs and charming beaches are snatched up either by covetous landlords or by speculators who appreciate the beauties of nature in the spirit of a money changer appraising a gold ingot'.[46] This idea is also primary in the degrowth philosophy, as it is based on the radical critique of the monetary value system.

According to Reclus, land use, i.e. the construction and transformation of landscape, expresses the rationality and codes which guide the actions on the environment by every people and society, so the aesthetic values are very significant in order to assess the environmental implications:

> By means of its fields and roads, by its dwellings and every manner of construction, by the way it arranges the trees and the landscape in

general, the populace expresses the character of its own ideals. If it really has a feeling for beauty, it will make nature more beautiful. If, on the other hand, the great mass of humanity remains as it is today, crude, egoistic and inauthentic, it will continue to mark the face of the earth with its wretched traces.[47]

For a fundamental change in humanity's relationship to nature, a revolution in values is certainly needed. But the ideological transformation that will result in the triumph of 'respect and feeling' can only succeed if there is a complementary process of social transformation, a change that would overturn the dominance of those 'industrial or mercantile interests'.[48] In fact, Reclus ought to be acknowledged in his historical scientific-intellectual context. He did not evade a utilitarian aim in the usage of natural resources and in the modification of the physical environment, but his position is sufficiently counterbalanced with a holistic and organic conception of nature. In his time, it would have been easier to be an unconditional partisan of human progress in its domination over biophysical laws. Yet he responded by advocating for a more integral perspective that meant a seismic change, i.e., 'a break with the dominant human-centered ideology'.[49] Reclus' ethical conception was based in a geocentric point of view rather than an anthropocentric one insofar as he considered human nature – both physiological and spiritual – as being deeply linked to its terrestrial condition. To Reclus, paraphrasing K. Ritter, 'the earth is the body of humanity and that man, in turn, is the soul of the earth'.[50]

Moreover, challenging the commodification and reification of nature, Reclus recognised an intrinsic value in its living and non-animated components, a conceptual prelude of a new universal ethics transcending human affairs, which integrates as moral subjects non-human life and even inert elements. According to Clark and Martin,

> Reclus therefore launches a scathing critique of humanity's abuse of the earth. In 'The Feeling for Nature' he wrote of the 'secret harmony' that exists between the earth and humanity, warning that when 'reckless societies allow themselves to meddle with that which creates the beauty of their domain, they always end up regretting it.[51]

Meaning of life and well-being

During the nineteenth century, a significant number of intellectuals showed their scepticism about the growing perfectibility and improvement of human beings,[52] and we may include Reclus among them. In Reclus' works, we can also find some assertions questioning growth as a condition for well-being. First and foremost, Reclus was aware of the polysemy of the term *progress*, which he analysed in depth. So, he perfectly referred to different interpretations that progress had acquired along the evolution of western thought and according to the historical times: from those who believe there are no boundaries and limits for human progress to those who feel symptoms of decadence in the course of history of civilisations. A similar issue may be found in degrowth scenarios: the apocalyptic view, which proclaims the inevitable environmental collapse; and the constructive view that advocates an alternative society, self-managed and self-limited. According to García both versions remind us of the respective connotation of two main figures of western thought[53]: Rousseau exemplifies the 'optimistic' version; and Hobbes, the 'pessimistic' one. Reclus seems to be closer to Rousseau's views than to Hobbes's: he dignified the progress achieved by humankind, but did not hesitate in counterbalancing with what he called 'regressions', or the things we have lost in every technical enhancement.

In particular, Reclus regretted the loss of primary values that characterised the primitive and pre-modern societies, which involved harmoniously into nature:

> The human being grows, but in the process he moves forward, thus losing part of the terrain that he formerly occupied. Ideally, civilised man should have kept the savage's strength, dexterity, coordination, natural good health, tranquillity, simplicity of life, closeness to the beasts of the field, and harmonious relationship to the earth and all beings that inhabit it.[54]

He praises the livelihood of certain communities, though they were not beneficiaries of the technical and scientific progress of *civilised* societies: 'Numerous cases can be found in which there is both moral superiority and a more serene appreciation of life among so-called savage or barbarous societies'.[55]

Moreover, 'regressions' also mean an attitude of returning to the past, aiming to appreciate the utility of old tools and techniques which have been shaded or detached by the rising irruption of new and advanced technologies. This attitude is also guided by an environmental sensibility:

> Old equipment, as well as men who are accustomed to a previous form of labour, are discarded as useless; however, the ideal is to know how to utilise everything, to employ refuse, waste, and slag, for everything is useful in the hands of one who know how to work with the materials.[56]

So, would *regress* be a synonym for degrowth? Similarities seem evident. For Latouche, degrowth is not 'negative growth': it is a new qualification of growth, or rather, it is different.[57] Likewise, Reclus emphasised the positive connotation of 'regress', in order to learn from traditional societies and their values. Perhaps, it might not be an absolute synonym of degrowth, but undoubtedly it fits very well within the degrowth proposal formulated by S. Latouche who believes in the interdependence of eight 'Rs' within a virtuous circle of autonomous quiet contraction, in order to create new economic relationships between wealthier and poorer countries: 're-evaluate, reconceptualise, restructure, redistribute, relocalise, reduce, re-use and recycle'.[58]

Degrowth has been largely discussed by considering the necessary diminution of the material goods we consume. Like environmentalism, it sees pre-modern cultures as magnificent examples of sustainable values, regarding their strong connection with the exploitation of primary resources, and livelihoods as perfectly fitted to natural events and paces. Nevertheless, as Latouche states: 'It would be unfair to rate degrowth partisans as technophobe and reactionary under the sole pretext they claim a "right to inventory" of progress and technique – a minimum demand for the exercise of civic responsibility'.[59] Like degrowth philosophers, Reclus was not nostalgic in the sense of believing that *any time in the past was better*. Rather than yearning for primitive human stages in history, he was passionate about the progress being undertaken in science and technology. Yet, his faith in scientific and technical progress had a clear pedagogical and prefigurative connotation, i.e. he hoped that one day it would enable man to dominate and exploit properly the natural environment, as a vast reserve of wealth and forces.[60] According to Clark and Martin, 'despite Reclus' statements concerning the greater degree of progress in modern societies, his writings demonstrate considerable sensitivity to the values and achievements of premodern and non-Western societies'.[61]

Notwithstanding, it is obvious that Reclus missed a set of values underlying human welfare that are seriously in decline, such as simplicity, tranquillity, spiritual connection to nature and its beings. He thought that the deterioration in modern and advanced societies affects not only a spiritual dimension (which would be enough to justify it) but also in terms of material conditions of life, as he pointed out, an impoverishment of environmental physical qualities. In any case, one could argue that he matched degrowth by advocating a simplification of life and a reduction of material consumption in order to achieve happiness.[62] His concern about the aesthetical values of the landscapes expressed a great sensibility to the qualitative aspects of well-being. Therefore, Reclus expressed the connection between these aesthetic values and certain dimensions of spirit: state of mind, happiness, etc., 'where the land has been defaced, where all poetry has disappeared from the countryside, the imagination is extinguished, the mind becomes impoverished, and routine and servility seize the soul, inclining it toward torpor and death'.[63] Reclus, according to his philosophy of nature, stressed that through spiritual involvement human beings might find their happiness and when they are sad, their regrets are at least mitigated by 'the sight of the wild countryside'.[64]

Bioeconomics

For degrowth partisans, the material contraction of the economic system is inevitable, due to the exhaustion of the bulk of global natural reserves, the disruption of thermal regulation systems and the surpassing of ecological boundaries with no point of return: peak oil, climate change, extinction of species, deterioration of fresh water, overexploitation of fisheries, etc. One determinant aim of degrowth philosophy is to achieve a sensible reduction in the material charge of human welfare, in order to make the productive system and consumption patterns more ecologically sustainable. Its first scientific foundations appeared in the 1960s and 1970s with the works of N. Georgescu-Roegen, who is acknowledged as the main intellectual source of the ecological approach of degrowth.[65] According to him, 'human activity transforms energy and materials of low entropy or good quality into waste and pollution which are unusable and have high entropy'.[66] Economic growth fails due to its optimistic and linear vision, as it cannot be disassociated from nature in terms of irreversible deterioration of matter and concentrated state of energy. Reminiscent works were the economic writings of Sergueï Podolinsky (1850–1891), Frederik Soddy (1877–1956) and the urban ecology of Patrick Geddes (1854–1932), authors who have usually been linked to anarchist theory and utopias. Thus, is Reclus a pioneer and an inspiration for the field of contemporary bioeconomics?

If there is something distinctive in Reclus' thought, it is the emergent, historical and dialectical view of human–nature relationships. Understanding humans as a constitutive part of nature avoids understanding 'humankind' and 'nature' as isolated and fully opposite. This is a basic premise for bringing the economic system closer to the *economics of nature*, as the former has to be analysed as a subsystem of the latter, both in evolutionary and physical terms. So, nature and humankind ought not to be separated in their joint evolution. Though he inherited a classical dualistic conception (with its roots in Aristotelian ontology), Reclus underlined the conceptive function of Mother Nature, viewing humans and societies as outcomes of the natural course of things. So, first and foremost, he was fully aware of our material dependence on nature in order to survive and progress:

> It is from her (the Earth) that we extract our materials; it is she who supports us with her nourishing juices and provides the air for our lungs; from a material point of view she gives us 'life, movement and being'.[67]

Such a vision remembers pre-modern societies, eastern traditions and animistic religions, which influenced anarchist groups,[68] and which are also evident cultural milestones for the theoretical framework of degrowth. Reclus was aware of the innate complexity in the reciprocal interactions between human societies and their environment. He 'long ago supported a more judicious and theoretically balanced dialectical view that avoids the extremes of overemphasising either order or chaos'.[69]

In the 1960s and 1970s, several exponents of radical environmentalism, such as R. Margalef, H.-T. Odum and B. Commoner, proposed that human systems (e.g. economic systems, urban systems), ought to learn functions and behaviours of ecosystems in order to become as efficient and clean as the latter. This idea is contained in the philosophy of 'biomimetics',[70] i.e. the imitation of nature as a means for reconstructing human productive systems in order to make them compatible with the Biosphere. It is difficult not to see a connection with the early anarchist geographers' thought, insofar as 'mutual aid'[71] understands nature as a source of inspiration for organising societies, but also focuses on its sociopolitical aspects. The implications of cooperation as the central axis of both natural and human societies is also guided by being more self-sufficient in conditions of lack of resources or environmental limitations, as Kropotkin attempted to show in his *Mutual Aid*. Both Élisée Reclus and Lev Metchnikoff, other contemporary anarchists, were involved in the elaboration of this work.[72]

In the same line of bioeconomics and biomimetics, it is quite conspicuous that Reclus sensed and proposed solutions that today are being fostered for reducing the generation of waste and recycling within the city. He trusted in the capacity of science and technology for designing integral management systems for polluted water. In this regard, he conceived the city as an 'organic system', which is analogous to the idea of 'societal metabolism', put forward by degrowth partisans.[73] Yet, the metabolism of cities may no longer be compared to the efficacy of an ecosystem in closing the cycles and being self-sufficient in the production and regeneration of matter. However, he advocated that cities have to learn the function of ecosystems and reintroduce waste outputs. He foresaw solutions closer to compost, suggesting that the urban waste water, once purified, might be transformed into manure for agriculture, through techniques not too sophisticated and innovated, but inspired by how nature avails of its own *waste*.[74]

Reclus understood, as Kropotkin did, that the best scenario for implementing balanced and harmonic strategies of good living were the rural communities, which is close to the idea of 'relocating' production and lifestyles in degrowth's virtuous circle.[75] In fact, relocation leads to a reduction of the size of economy, according to the principles of biomimetics. Early anarchist geographers saw traditional rural communities like Russian *mirs* as units of production and consumption, where every member would get from the commons what they needed, independently of what they had produced.[76]

Following these insights, Reclus advocated strategies of self-limitation as the best way to use and manage the land, an idea undoubtedly in line with degrowth philosophy. In this regard, he exhorted peasants to produce according to their individual needs, but without compromising the needs of other workers:"

> The amount of land to which the individual, the family, or the community of friends has a natural right is the amount that can be worked through individual or collective labour. As soon as a parcel of land exceeds the amount that they are able to cultivate, they would be wrong to claim this additional portion. Its use belongs to another worker.[77].

In sum, Reclus' thought is full of arguments in favour of sufficiency and voluntary simplification as a strategy of 'good living' and to constitute a more sensible economy in order to meet the basic needs of humans.[78]

Democracy

In physical terms, and transcending any ideological or political connotation, degrowth seems to be the unique path to reduce the pressure of capitalist and consumerist lifestyles over the global ecological boundaries. In particular contexts, capitalism and economic growth may convince us of eternal abundance, through different discourses: dematerialisation, weak sustainability, green technologies… but in overall terms this economic logic inevitably will be subjected to the imperative of entropy and non-reversible degradation.

In this regard, one can distinguish two possible scenarios of how this transition may be implemented[79]: i) a self-managed material degrowth, which is implemented through participatory and voluntary processes; ii) an imposed degrowth through the rise of *eco-dictatorships* and the proliferation of conflicts and struggles motivated by the anxious extraction and usage of increasingly scarce resources.[80] As Deriu asserts,

> while a perspective of material degrowth is possible even in an authoritative manner, a political philosophy of degrowth requires a more coherent reflection on a change of the ideas of well-being or *buen vivir*, on the transformation of the economic rules and on the renewal of the institutions of democracy.[81]

Reclus' ideal society had no state, bureaucracy and legislation to be used for domination and subjugation. In this regard, Reclus thought that 'respect for human laws in disregard for the higher moral law is no virtue and indeed amounts to no more than "moral cowardice"'.[82] He exemplified this concept in the following way:

> there will be only brothers who have their share of daily bread, who have equal rights, and who coexist in peace and heartfelt unity that comes not out of obedience to law, which is always accompanied by dreadful threats, but rather from mutual respect for the interest of all, and from the scientific study of natural laws.[83]

In addition, he exhorted citizens not to vote as to do so 'is to give up your own power…, to put on others' shoulders the responsibility of one's actions is cowardice.'[84] This opens an interesting discussion around 'growth' and 'development' and its colonising effect, as they are presented as legitimated aims within the political and institutional arena. They become tools of control and domination of the imaginary, and even more a way to justify the hegemony of industrial societies over the developing and poorer ones. Is an anarchist political conception compatible with degrowth aims?

Degrowth seeks to guarantee a set of political requirements to be implemented by means of a voluntary transition in fields such as consumption, work and the universe of values.[85] In a nutshell, degrowth partisans are convinced that a monitored and guided degrowth has to be based on *bottom-up* strategies rather than the *top-down* practices which have characterised the policies for 'sustainability' by both capitalist and communist regimes. But, is this a call for an abolition of state and bureaucracy? On this point, there can be conflicting positions about preserving to a greater or lesser extent the political institutions linked to the capitalist state. Thus, on the one hand, some of degrowth's partisans defend the current democratic institutions 'considering the risks of losing what we have achieved'.[86] On the other hand, a radical strand 'demand(s) completely new institutions based on direct and participatory democracy (more alternative, or post-capitalist vision)'.[87] As degrowth, anarchism would trace different routes to a more sustainable society, insofar as to give a determinant role to the citizenry which should involve active participation in political life. Thus, degrowth challenges the centralised, hierarchised and representative democracy and needs to be thought in 'a broad and articulated process of shared learning, self-education, reconstruction of social ties and collective transformation'.[88]

Following Reclus, a self-managed degrowth should reduce the prominence of the State and its large influence in social and individual life. The French geographer considers the State as an obstacle given its hierarchical and complex structure which leads to a wastefulness of resources and intrinsic inefficiency. He believed that

> through the phenomena of human activity in the arenas of labour, agriculture, industry, commerce, study, education, and discovery that subjugated peoples gradually succeed in liberating themselves and in gaining complete possession of that individual initiative without which no progress can ever take place.[89]

Indeed, this approach may help to challenge diverse conflicts (such as environmental ones) which concern different nations at international scale, as Reclus suggested:

> Each accomplishment that is thus realised without the intervention of official bosses and outside the state, whose cumbersome machinery and obsolete practices do not lend themselves to the normal course of life, is an example that can be used for larger undertakings.[90]

Unlike other anarchist thinkers of his day, Reclus did not develop ideal societies and utopias based on anarchist principles. His reflection had a broader character and according to this overall view, every individual has to ensure the basic conditions for life, material and non-material needs, for achieving the full development of every human. This standpoint is in line with the Aristotelian *telos*, as the achievement and full realisation of self. Reclus and Kropotkin 'assumed that sustainable human/environment relations could only be initiated through social transformation and fundamental changes in human values that would promote the demise of capitalism, racism, the modern State, gender inequities, and other forms of social hierarchy'.[91]

Though Reclus trusted in the role of science and technology for driving an evolution and transformation to autonomous and dis-aliened human societies, he also acknowledged that they may be used for evil purposes. Herein it is possible to find a point of connection with the so-called 'post-normal science',[92] which has been attached to the debates around degrowth transition.[93] The voice of expertise in decision-making around conflicts and risky challenges on issues such as environmental problems, health, social affairs, has prevailed over popular wisdom or the perception of people. A more participative involvement of the population in a context of degrowth, full of uncertainty and systemic shifts, demands an equal treatment of these voices, along with the technical and political ones.

Yet, Reclus was aware that, by means of communal work, it is easier to achieve a balanced relation with nature and full sense of good living. In this regard, he saw political institutions and rich classes as the enemies of preserving these values:

> Thus the spirit of full association has by no means disappeared in the communes, despite all the bad will of the rich and the state, who have every interest in breaking apart these tightly bound bundles of resistance to their greed or power and who attempt to reduce society to a collection of isolated individuals.[94]

This feeling of community is based on intercultural approaches: 'Traditional mutual aid occurs even among people of different languages and nations'.[95] Such complementarity between livelihood and cooperativism is similar to the cross-cultural conception of degrowth, advocated by post-development thinkers and based on the philosophy of *buen vivir*.

Justice

Degrowth also seeks to dismantle the idea that economic growth will release the so-called developing countries from poverty, under the eternal promise of 'development'. In fact, this mantra is joined to the origin of the term 'under-development'. From the mid-twentieth century, the idea of development was a key determinant in the process of political independence for most of the African and Asian colonies and the creation of new states.[96] Social imbalances have progressively been increasing, despite 'one common assumption among economists [which] is that only economic growth can improve the living conditions of poor people on the planet'.[97]

The critical reaction of degrowth to this axiom is that economic growth is unfair, centralised, and no longer reduces the social and economic rifts in terms of wealth and welfare among countries, but also among different social strata. At a global scale, inequalities are laid on: i) the historical 'ecological debt' of rich countries to poor countries; and ii) the market as a system which favourably compensates manufacturing and trade above primary production or extraction, leaving in a worse position those countries whose economies are based on agriculture, raw materials or crude fuels. Obviously, this critique is quite a lot more complex, as it also questions who manages these resources, how land is distributed and tenured, etc. But it is very similar to the ethos of Reclus' thought. For him, social injustice and environmental damages are two sides of the moral and material deterioration of humankind. Reclus' political thought and activism is thus strongly connected to his environmentalism, insofar as no separation exists between human exploitation of nature and human domination and exclusion over other humans in his insights. Further-more, learning about nature is needed in order to define a new social and political archetype. He argued that only a restoration of the harmonious relation between humans and nature would lead to a world free of injustice and widespread socioeconomic imbalances.[98]

This connection between social injustice and environmental exploitation seems to be akin to that of 'environmental justice', which is one of the strands cultivated by degrowth thinkers.[99] However, it would be risky to state that Reclus was thinking in terms of 'environmental justice' as it is understood today. Environmental justice has become a demand over the world, but originally it was linked to the expression 'environmentalism of the poor',[100] within the studies of political ecology. This idea refers to Global South communities (mostly indigenous people) and their claims versus extractivism, land confisca-tion, pollution and environmental degradation fostered by foreign corporations, national policies and their chrematistic aims. Certainly, Reclus referred to the unequal and unfair distribution of land and means of production within society. So, for Reclus, workers and peasants, once they achieve the control of these means and the instauration of the commons, will lead society to a closer and more peaceful involvement with nature.

Yet, according to the environmental justice approach, it is worth noting that Reclus was one of the few thinkers in his times opposed to colonisation and imperialism, i.e. a proto-critical-thinker of social injustice at a global scale. F. Ferretti confirms this statement,[101] challenging authors who put for-ward the hypothesis that Reclus was a 'colonialist', or that he was less critical of French colonialism in Algeria than of British colonialism in India. This was mainly a misunderstanding, due to the differentiation of two kind of colonialism: i) settlement colonies; and ii) invaded colonies.[102]

Reclus approached the latter critically, anticipating much of the post-colonial studies and even post-development critical theory, from which several trends of degrowth have emerged. He was particularly critical of the French coloni-alism in Africa and Asia, questioning whether exploitation, slavery and impoverishment of colonies and their people has to be the price to pay to consolidate the idea of a more prosperous Nation-State for the French citi-zenship.[103] Thereby, Reclus was arguing not only in terms of colonialism as political control but also in terms of 'ecological plundering' carried out by the great political powers of Europe and North America upon their colonial periphery. He witnessed the early stages of the depletion of natural resources and environmental degradation connected to industrialisation, and how the European countries had to turn to vast lands of Africa and Asia in search of raw materials given the exhaustion of their own reserves. Moreover, he foresaw new strategies of colonialism, e.g. the economic and political interest of the USA over the newly independent countries of South America, 'whether by appealing invitations or masked orders or making them feel the weight of its high protection'.[104] Then, he quite rightly visualises the complexity of geopolitics that were then shaping new forms of colonialism, the most determinant factor being the colonisation of the imaginary with economic aims (growth, development, incomes), as degrowth reveals today.

In this regard, 'decolonising the imaginary' means not only restructuring our way of thinking about economy, development and human welfare, but also by opening up a dialogue among different approaches and philosophies of welfare and good living.[105] There is a western cultural stigma in the origin and spreading of this economic imaginary that has hidden alternatives and traditional styles of material and spiritual welfare. Reclus' thought points in such a direction, and he refers to the issue of justice in order to achieve a more balanced relation with nature. According to F. Ferretti, 'he represents the principle of unity of human kind, stating that all men should live on the Earth as "brothers" and refusing the "superiority" of one culture over the others'.[106] Thus, Reclus rejected racism, which is an obstacle in the transition to degrowth, as Latouche warns: 'we resist, and must resist all forms of racism and discrimination (skin color, sex, religion, ethnicity)' which are 'all too common in the West today'.[107] He questioned progress if based on centralisation, hierarchy, specialisation, privilege and domination[108] and he strongly challenged those who use it for justifying injustice and oppression in the name of progress.[109] Regarding this idea, Reclus wondered: 'what kind of "progress" is it for the people of Cameroon and of Togo to have henceforth the honour of being protected by the German flag, or for the Algerian Arabs to drink aperitifs and express themselves elegantly in Parisian slang?'[110]

In addition, Reclus acknowledged the capacity and the right of self-organisation of the colonies by their natives. Though he was a firm supporter of human unity, for him this must be reached through a non-hierarchised way, without judgments of superiority or inferiority and 'would bring Europe and its Others increasingly closer to each other materially and culturally'.[111] The recognition of otherness with respect to the centralised power of Europe in the world is intimately connected with the advocacy of degrowth for establishing a dialogue among different philosophies of livelihood, transcending and criticising the monist way of understanding welfare. In short, for Reclus, 'it is anarchist morality that is most in accord with the modern conception of justice and goodness'.[112]

Final remarks: summarizing the contributions of Reclus for degrowth ethos

Throughout this chapter, it has been shown that many writings and works of Élisée Reclus address essential issues which are today at the core of degrowth theory. His radical thought has obvious bonds with this revolutionary view of human welfare: a critical view of progress and its instrumental aims; the concern about human environmental implications and the degradation of nature; the idealisation of local, self-managed, emancipated and balanced communities with environmental qualities; the denouncing of injustice between the rich and the poor at diverse scales; and the appreciation of primitive and traditional cultures as references for good living.

Moreover, Reclus' 'social ecology' provides degrowth determinant foundations that may dismantle the reductionist understanding of nature fed by orthodox economics. He appreciated the importance of natural causalities on human progress and the viability of civilisations. The result is a moderate deterministic discourse which is necessary for the cultural challenges that the degrowth transition demands. Reclus understood human systems in comparison with natural systems. He referred to the 'organic system' which is very close to the idea of 'metabolism', a key word in the ecological approaches of degrowth, bioeconomics, urban ecology and ecological economics. Also, he offered a dialectic and diachronic assessment of scientific, material and technical evolution of civilisations and a historical deconstruction of the discourse of 'progress', which leads to and is at the core of the contemporary reductionist idea of development. Thus, the 'decolonisation of the imaginary' ought to start with a 'deconstruction' of the roots of the current economic system, and Reclus' insights in 'The Progress of Mankind' (1896) are definitely a milestone in the historical-critical analysis of this discourse.

Reclus introduced the aesthetic dimension in order to avoid a utilitarian conception of nature, and criticises the commodification of and speculation over natural and cultural qualities of the landscape. Degrowth proposes a reduction in the material charge of economy. Using the moral insights of Reclus as inspiration, this contraction may emerge in a *requalification* of human well-being, i.e. acknowledging the spiritual and aesthetic values of the environment. Complementary to this, Reclus' worldview of humans, and their position on the Earth, provides degrowth with foundations for a new imaginary that replaces the utilitarian and mercantilist conception of good living: economic relations based on commons, solidarity and justice; environmental relations based on sufficiency, knowledge of ecological boundaries, and broad sense of respect for and solidarity with non-human life. Reclus' works such as *Histoire d'un ruisseau* (1869) and *Histoire d'une montagne* (1880), alongside many essays and writings, are worthy of being among these references, in which he offered magnificent insights of phenomenological immersion in the goodness and equilibrium transmitted by a beautiful landscape or vigorous geographic features.[113]

A significant part of the 'degrowth' discourse in Reclus has to do with, on one hand, his facet as a geographer. As contemporary holistic thinkers, his theory handles basic philosophical and ontological categories. Yet, the dialectical treatment of determinant aspects of humankind and their evolution, such as progress, technology, science, culture, etc., distinguishes him from other contemporary thinkers, by offering contrasted and fully argued approaches that bring him closer to the complexity of these issues. This holistic and cross-disciplinary approach is fundamental to the elaboration of a complex framework of arguments and insights which are very useful for the contemporary challenges opened up by degrowth trends. He relativised and diversified the idea of environment (milieu), understanding it as a social construction and an emergent and coevolutionary process between humans and their physical surroundings.

On the other hand, although characterised by what we might call today an eco-anarchist approach, social commitment, dissidence and a transformative and revolutionary spirit, are required conditions for a degrowth transition. Reclus' political thought opens interesting discussions around the abolition of State, the contradictions of democracy, the role of education in alienating children (in the same line as Ivan Illich, who proposed the abolition of official education) and the configuration of decentralised communal organisations to manage political life more effectively than central and hierarchised bureaucracy. Material reduction of economics, according to degrowth, implies relocalisation of productive and consumptive patterns. Though he did not refer explicitly to *utopia*, an ideal society for him is based on small communities and an integrative and communal conception of work, production and life, satisfying basic needs of every member. Unlike the radical sector of degrowth, Reclus praised and showed the considerable improvement that humankind had achieved enhancing material well-being, in areas such as agriculture, communications and the mechanisation of production work. Yet, he conceived of this improvement as guided by non-capitalist values. Efficiency and productiveness only have real presence within non-competitive and non-lucrative principles.

In sum, I am arguing that Élisée Reclus should be a reference for contemporary thinkers of degrowth, as determinant as the various authors (and others) referred to in this chapter. As mentioned in the first section of this chapter, degrowth is a slogan itself, an iconic term devoted to rethinking our role and fate as humans in our home, the Earth. Reclus condensed perfectly all this into what is probably his most significant quote: 'Man is Nature becoming self-conscious'.[114] In this respect, it is worth concluding with the words of Clark and Martin, explaining the meaning and scope of this assertion:

> This concept... captures the essence of Reclus' message: that humanity must come to understand its identity as the self-consciousness of the earth and that it must complete the process of developing this consciousness in history. In effect, he proposes a theoretical project of understanding more

fully our place in nature and of unmasking the ideologies that distort it, and a corresponding ethical project of assuming, through a transformed social practice, the far-reaching moral responsibilities implied by that crucial position.[115]

There is no better way to depict what a transition to self-managed degrowth societies means.

Notes

1 Latouche, *Survivre au développement*; 'Decolonization of Imaginary', in d'Alisa, Demaria and Kallis (eds), *Degrowth: A Vocabulary for a New Era*.
2 Latouche, *Le pari de la décroissance; Farewell to Growth*; Martínez-Alier et al., 'Sustainable De-growth: Mapping the Context, Criticisms and Future Prospects of an Emergent Paradigm'; Kallis and March, 'Imaginaries of Hope: The Utopianism of Degrowth'.
3 Escobar, 'Degrowth, Postdevelopment, and Transition: A Preliminary Conversation'.
4 Flipo, 'Voyage dans la galaxie décroissante'; Bayon, Flipo and Schneider, *La Décroissance, 10 questions pour comprendre et en débattre*; Demaria et al., 'What is Degrowth? From an Activist Slogan to a Social Movement'.
5 Sippel, 'Back to the Future: Today's and Tomorrow's Politics of Degrowth Economics (Décroissance) in Light of the Debate over Luxury among Eighteenth and Early Nineteenth Century Utopists'.
6 Sippel, 'Back to the Future', 26.
7 Clark and Martin, *Anarchy, Geography, Modernity: Selected Writings of Elisée Reclus*, 10.
8 Gómez, Muñoz and Cantero, *El pensamiento geográfico: estudio interpretativo y antología de textos: de Humboldt a las tendencias radicales*.
9 Giblin, 'Reclus: un écologiste avant l'heure?'; Marshall, *Demanding the Impossible: A History of Anarchism*; Clark and Martin, 'Anarchy, Geography, Modernity'; Springer, 'Anarchism and Geography'.
10 Giblin, *Élisée Reclus. El Hombre y el Tierra*, 62, trans. F. Toro.
11 Giblin, 'Reclus: un écologiste avant l'heure?'
12 Springer, 'Anarchism and Geography. A Brief Genealogy of Anarchist Geographies', 49–50.
13 Latouche, 'Can the Left Escape Economism', 75.
14 Clark and Martin, *Anarchy, Geography, Modernity*, 17.
15 Ward, *Anarchism: A Very Short Introduction*, 90; Clark and Martin, *Anarchy, Geography, Modernity*, 16; White and Kossoff, 'Anarchism, Libertarianism and Environmentalism: Anti-Authoritarian Thought and the Search for Self-Organizing Societies', in Pretty et al. (eds), *The SAGE Handbook of Environment and Society*.
16 Toro, 'Educating for Earth Consciousness. Ecopedagogy within Early Anarchist Geography', in Springer, de Souza and White (eds), *The Radicalization of Pedagogy. Anarchism, Geography, and the Spirit of Revolt*.
17 Giblin, 'Reclus: un écologiste avant l'heure?'.
18 Clark and Martin, *Anarchy, Geography, Modernity*.
19 A low estimation of Reclus' environmental sensibility may be exemplified in the most representative anarchist thinker of social ecology in the twentieth century, Murray Bookchin ('Deep Ecology, Anarcho Syndicalism, and the Future of Anarchist Thought.' in Bookchin et al. (eds), *Deep Ecology & Anarchism. A Polemic*', 37). He strongly criticises (even in a risible way) the spirituality with which Reclus sees harmony between humans and nature:

[Reclus'] view that 'secret harmony exists between the earth and people', one that 'imprudent societies' will always regret if they violate it, is far too vague, at times even mystical, to be regarded as more than a generous sentiment... I would certainly praise Reclus as an anarchist and a resolute revolutionary, but I would be disquieted if his particular views on the natural world were identified, apart from their good intentions, with eco-anarchism.

20 Demaria et al., 'What is Degrowth?', 196.
21 Purchase, *Anarchism and Environmental Survival*, 5.
22 Ferretti and Pelletier, 'En los orígenes de la geografía crítica. Espacialidades y relaciones de dominio en la obra de los geógrafos anarquistas Reclus, Kropotkin y Mechnikov', 4.
23 Pelletier, 'Élisée Reclus et la mésologie'.
24 Reclus, *L'Homme et la Terre*, vol. 1, 39.
25 Reclus, *L'Homme et la Terre*, vol. 1, 81.
26 Clark and Martin, *Anarchy, Geography, Modernity*, 25.
27 Pelletier, 'Élisée Reclus et la mésologie'.
28 Pelletier, 'Élisée Reclus et la mésologie'.
29 Reclus, *L'Homme et la Terre*, vol. 1, IV.
30 Reclus, *The Ocean, Atmosphere and Life*, 522.
31 Reclus, 'L'homme et la nature: de l'action humaine sur la géographie physique', 763.
32 Reclus, 'Du sentiment de la nature dans les sociétés modernes', 379 [hereafter 'Du sentiment'].
33 Ehrlich, *The Population Bomb*.
34 Martínez-Alier, 'Neo-Malthusians', in d'Alisa, Demaria and Kallis, *Degrowth*, 125.
35 Martínez-Alier, 'Neo-Malthusians', 125.
36 Kallis, Demaria and d'Alisa 'Introduction: Degrowth', in d'Alisa, Demaria and Kallis, *Degrowth*, 9.
37 Masjuan, *La ecología humana en el anarquismo ibérico: urbanismo orgánico o ecológico, neomalthusianismo y naturismo social*.
38 Reclus, *L'Homme et la Terre*, vol. 5, 418.
39 Reclus, *L'évolution, la révolution et l'idéal anarchique*, 135–136.
40 Castoriadis, *The Imaginary Institution of Society*; Latouche, *Survivre au développement; Farewell to Growth*; 'Decolonization of Imaginary'.
41 Latouche, *Survivre au développement*.
42 Sbert, 'Progress', in Sachs (ed.), The Development Dictionary. A Guide to Knowledge as Power. Reclus, at the same time, criticises and advocates the idea of progress. This facet of his thought seems to be contradictory. Yet, such a position is rather very close to post-development thinkers today: 'He recognizes that since the French Revolution the idea of progress has often been used as an ideological justification for elitism, class domination, imperialism, and other evils' (Clark and Martin, *Anarchy, Geography, Modernity*, 35). Thus, in line with Sbert, development is nothing but a continuation of the same colonising effect.
43 Reclus, *L'Homme et la Terre*, vol. 6, 257.
44 Reclus, 'L'homme et la nature', 766.
45 Clark and Martin, *Anarchy, Geography, Modernity*, 39.
46 Reclus, 'Du sentiment', 377.
47 Reclus, *Histoire d'une montagne*, 299–300.
48 Reclus, *The Ocean, Atmosphere and Life*, 526.
49 Rothen, 'Élisée Reclus' Optimism', in Ishill (ed.), *Elisée and Elie Reclus: In Memoriam*.
50 Reclus, 'L'homme et la nature', 762.
51 Clark and Martin, *Anarchy, Geography, Modernity*, 25.
52 Nisbet, *History of the Idea of Progress*.
53 García, 'Degrowth, the Past, the Future, and the Human Nature'.

54 Reclus, 'La grande famille', 8.
55 Reclus, *L'Homme et la Terre*, vol. 6, 512.
56 Reclus, *L'Homme et la Terre*, vol. 6, 531.
57 Kallis et al., 'Introduction', 3; Asara et al., 'Socially Sustainable Degrowth as a Social-Ecological Transformation: Repoliticizing Sustainability', 377.
58 Latouche, *Farewell to Growth*.
59 Latouche, *Le pari de la décroissance*.
60 Giblin, *Élisée Reclus*, 82.
61 Clark and Martin, *Anarchy, Geography, Modernity*, 38.
62 Ridoux, *La décroissance pour tous*; Latouche, *Survivre au développement*; Latouche, *Farewell to Growth*.
63 Reclus, 'Du sentiment', 379–380.
64 Reclus, 'Du sentiment', 380.
65 Latouche, 'Décroissance', *Farewell to Growth*; Martínez-Alier et al., 'Sustainable De-growth'; García, 'Degrowth'.
66 Cited in Demaria et al., 'What is Degrowth?', 198.
67 Reclus, 'L'homme et la nature', 762.
68 Marshall, *Demanding the Impossible*; Rapp, *Daoism and Anarchism: Critiques of State Autonomy in Ancient and Modern China*; Springer, 'War and Pieces'.
69 Clark and Martin, *Anarchy, Geography, Modernity*, 20.
70 Riechmann, *Biomímesis: ensayos sobre imitación de la naturaleza, ecosocialismo y autocontención*.
71 Kropotkin, *Mutual Aid: A Factor of Evolution*.
72 Ferretti and Pelletier, 'En los orígenes'.
73 Şorman, 'Societal Metabolism', in D'Alisa, Demaria and Kallis (eds), *Degrowth: A Vocabulary for a New Era*.
74 Toro, 'La sensibilidad ecológica en el pensamiento de Elisée Reclus: una revisión necesaria para entender las claves de la crisis ecológica contemporánea'.
75 Latouche, *Farewell to Growth*.
76 Reclus, 'A mon frère le paysan'.
77 Reclus, 'A mon frère le paysan', 5.
78 A philosophy that moves in the same line as other contemporary environmentalists, such as Henry David Thoreau.
79 García, 'Degrowth'.
80 Sempere, 'Decrecimiento y autocontención'; Toro, 'Una visión crítica de la sostenibilidad y algunas reflexiones para salir del imaginario dominante', in Farrés and Matarán (eds), *Otro municipio es posible? Guanabacoa en La Habana*.
81 Deriu, 'Democracies with a Future: Degrowth and the Democratic Transition', 560.
82 Cited in Clark and Martin, *Anarchy, Geography, Modernity*, 52.
83 Reclus, 'L'Anarchie', 7–8.
84 Reclus, *Correspondencia 1850–1905*.
85 Latouche, *Farewell to Growth*; Kallis et al., 'Introduction'; Trainer, 'The Degrowth Movement from the Perspective of the Simpler Way'.
86 Demaria et al., 'What is Degrowth', 199.
87 Demaria et al., 'What is Degrowth', 199.
88 Deriu, 'Democracies with a Future', 560.
89 Reclus, *L'Homme et la Terre*, vol. 6, 223.
90 Reclus, *L'Homme et la Terre*, vol. 6, 223.
91 Breitbart, 'Anarchism/Anarchist Geography', in Kitchin and Thrift (eds), *International Encyclopaedia of Human Geography*, 109.
92 Funtowictz and Ravetz, 'Science for the Post-Normal Age'.
93 D'Alisa and Kallis, 'Post-normal Science', in d'Alisa, Demaria and Kallis, *Degrowth*.

94   Reclus, *L'Homme et la Terre*, vol. 6, 270.
95   Reclus, *L'Homme et la Terre*, vol. 6, 270.
96   De Rivero, *The Myth of Development: The Non-Viable Economies of the 21st Century*.
97   Demaria et al., 'What is Degrowth', 199.
98   Gómez, Muñoz and Cantero, *El pensamiento geográfico*, 44.
99   See, for instance, Anguelovski, 'Environmental Justice', in d'Alisa, Demaria and Kallis, *Degrowth*.
100 Martínez-Alier, *The Environmentalism of the Poor*.
101 Ferretti, '"They Have the Right to Throw Us Out": Élisée Reclus' *New Universal Geography*'.
102 Llorente, 'Colonialismo y geografía en España en el último cuarto del siglo XIX. Auge y descrédito de la geografía colonial'; Ferretti, 'They Have the Right to Throw Us Out'.
103 Giblin, *Élisée Reclus*.
104 Reclus, *L'Homme et la Terre*, vol. 6, 81.
105 Escobar, 'Degrowth, Postdevelopment'.
106 Ferretti, 'They Have the Right to Throw Us Out', 1337.
107 Latouche, *Farewell to Growth*, 56–57.
108 Olwig, 'Historical Geography and the Society/Nature "Problematic": The Perspective of J. F. Schow, G. P. Marsh and E. Reclus', 40.
109 Clark and Martin, *Anarchy, Geography, Modernity*, 35.
110 Reclus, *L'Homme et la Terre*, vol. 6, 504.
111 Ferretti, 'They Have the Right to Throw Us Out', 1351.
112 Reclus, 'L'Anarchie', 11.
113 Toro, 'Educating for Earth Consciousness'.
114 Reclus, *L'Homme et la Terre*, vol. 1, 1.
115 Clark and Martin, *Anarchy, Geography, Modernity*, 17.

6 Revolutions and their places

The anarchist geographers and the problem of nationalities in the Age of Empire (1875–1914)

Federico Ferretti

Introduction: anarchism, nationalism and anti-colonialism

This chapter addresses the problem of nationalities in the work of early anarchist geographers at the Age of Empire.[1] Drawing on recent literature on anarchist geographies and histories of transnational anarchism, I address the work of three key exponents of the international network of the anarchist geographers, Mikhail Dragomanov or Drahomanov (1841–1895), Pëtr Kropotkin (1842–1921) and Elisée Reclus (1830–1905). My main argument is that the anarchist tradition and the idea of nation stand not in opposition, but in mutual relation, and that such relation was linked to early anti-colonialism in the Age of Empire. The standpoint of anarchist geographers is a privileged one to understand the anarchist idea of nation because these militants and scholars worked on territories, regions, and borders, thus the definition of what a nation is (or should be) was part of both their professional duties and their political interests.

Recent literature has analysed the works and networks of Dragomanov, Kropotkin and Reclus and of some or their fellows like Lev Mečnikov/Léon Metchnikoff (1838–1888); Charles Perron (1837–1909), Gustave Lefrançais (1826–1901) and others, in the context of their networks and their common scientific and militant connections. In particular, new scholarship stresses the decisive importance of a place like Switzerland for the acquaintance of these authors, who met there during their common exile in the 1870s and 1880s, when they worked together for Reclus's *New Universal Geography* and for the 'anti-authoritarian international',[2] then represented in Switzerland by the first anarchist organisation in history, the *Fédération jurassiene* (Jura Federation).[3] Their network is a very meaningful case study for at least two recent lines of research in international literature. The first one addresses anarchist geographies and their genealogies,[4] showing that geography and anarchism have a long tradition of relations and still share a number of contents toward a project of libertarian and egalitarian transformation of society and space. The second one deals with historical studies on transnational anarchism, which have shown that anarchism was firstly a transnational movement, since the Age of Empire, because its militants, more than any others, circulated throughout the world as political exiles, economic migrants, or committed international propagandists.[5]

If anarchism is first and foremost internationalist,[6] a transnational approach allows considering that the multilingual and cosmopolitan circulation of anarchist ideas and militants did not impede that internationalism, that the movement's hubs were situated, and that forms of cultural rootedness were not incompatible with migration and international solidarity. On the one hand, anarchist internationalism is traditionally opposed to nationalism, militarism and xenophobia. On the other hand, as Davide Turcato argues, 'an inclusive idea of nation does not clash with anarchism'.[7] To understand this point, it is worth considering first that anarchist federalism dissociated the idea of state from the idea of nation since the times of Pierre-Joseph Proudhon (1809–1865), who considered nations as spontaneous and bottom-up formations contrasting with the political divisions of Europe at that time.[8] Second, nineteenth-century European anarchism had its roots in the republican movements for social liberation which animated the revolts of 1848–1849,[9] and its militants were often acquainted with other activists struggling for national liberation in Eastern and Southern Europe: early anarchists generally supported these movements, hoping that national liberations could hasten social revolution on an international scale.[10]

Examples of these trends can be found in the links between anarchist historical figures like Mikhail Bakunin (1814–1876) and Errico Malatesta (1853–1932) and the Italian movement of the *Risorgimento*, including the republican training of the first generation of Italian anarchists, who broke with Giuseppe Mazzini after the Paris Commune, taking more radical positions. As Turcato points out,

> Mazzinian Republicanism is indeed a fundamental term of reference in discussing the beginning of the International in Italy ... At the same time, in Italy the International arose from the moral and intellectual discomfort and dissatisfaction of the idealist Italian youth toward Mazzinian republicanism.[11]

In this context, Pier Carlo Masini has analysed the 'heresies of *Risorgimento*', i.e. the radical and federalist tendencies which characterised Italian 'patriotic' movements during the nineteenth century,[12] while Arthur Lehning has highlighted the deception of the most radical militants after the monarchist and centralist turn that Italian unification took after 1861, arguing that this disappointment was decisive in inspiring Bakunin's radicalisation during his stay in Italy in the 1860s.[13] The 'Herodotus of Anarchism', Max Nettlau,[14] made a comparison between the revolutionary situation of Italy during the 1859 uprisings and the Spanish experience of the *Sexénio Democrático* (1868–1874), a movement which was supported by Bakunin and by future anarchist geographers like Metchnikoff and the Reclus brothers, Elie and Elisée.[15]

About France, Gaetano Manfredonia has analysed the complex relations between anarchism and republicanism, and their respective representations of the 1789 Revolution.[16] There, the importance of places and local identities for analysing social struggles was already clear to Bakunin: in his claims for supporting the French republican turn during the 1870 Franco-Prussian war, the Russian revolutionary argued that, for understanding the struggles of the French proletariat, it was worth considering its mentality and its identity. These identities were situated and rooted at a more complex scale than those of states and nations: in his work *Lettres à un Français* (1870), Bakunin quoted the feelings of Parisian revolutionaries, who expressed then concepts like: 'Prussian soldiers, you can defeat Napoleon III and put your flag on the Tuileries Palace: we leave you Notre-Dame and the Louvre, but you will never conquer this narrow and dirty street in Belleville.'[17]

It is also worth considering that anarchists' commitment to national liberation inspired their early anti-colonialist consciousness: recent studies consider anarchists as the first socialists of European origin to be acquainted with indigenous movements and non-European cultures: it is the case of the writings collected by Hirsch and van der Walt[18] and of the studies on 'anti-colonial imagination' by Benedict Anderson.[19] Recent studies on historical anarchist geographies have also highlighted the early and radical positions of scholars like Reclus in challenging colonialism, racism, and euro-centrism and in questioning the positivist idea of linear progress through 'scientific' arguments.[20] Recent works on anarchism and Indian de-colonisation by Maia Ramnath address the history of anarchists' commitment to Indian anti-colonialism (including not only Indian militants, but also British ones such as Guy Aldred) as an example of existing alternatives to experiences of decolonisation drawing on nationalism and traditionalism.[21] In his works on Japanese anarchism, Sho Konishi even argues that exchanges between Russian and Japanese radicals at the time of the Meiji Revolution (1868), inaugurated by the aforementioned Metchnikoff, inspired 'a transnationally formulated temporality and corresponding order of knowledge and practice that [Konishi calls] cooperatist anarchist modernity ... beyond western modern constructs'.[22]

The support anarchists gave then to struggles of national liberation in Eastern Europe, and especially in Poland, Ukraine, Greece and Finland (and for Western Europe in Ireland), can be read as an early critique of what Antonio Gramsci, and then Michael Hechter, called 'internal colonialism'[23]. Finally, the proximity between anarchism and radical republican movements can be explained by their general sharing of the idea of 'freedom per non domination' which, according to Philipp Pettit, characterises radical movements for emancipation, as this kind of freedom means the complete liberation of every individual from the discretional power of any other person.[24]

In the first part of the chapter, I present Dragomanov's geographical proposals for Ukrainian national liberation; in the second part, I address Kropotkin's ideas on Finland as a 'rising nationality'; in the third part, I present Elisée Reclus's relations with a South-American republic decolonized during the nineteenth century, Brazil.

Mikhail Dragomanov, 'a Ukrainian socialist-federalist'

The Ukrainian intellectual Mikhail Petrovič Dragomanov is considered today as a sort of Founding Father of the Ukrainian state, apparently forgetting that during his stay in Geneva from 1876 to 1889 he was acquainted with the anarchists, collaborated with Reclus for his geography, edited the correspondences of Bakunin in 1896[25] and claimed Proudhon's legacy. Nevertheless, it is true that Dragomanov was uncomfortable with the Russian centralism of a great part of the other exiles from Eastern Europe who animated the stubborn Slavic community in Geneva.[26] He declared all this in a pamphlet that he published in French to call the attention of Western public opinion to Ukraine's social and national problems.

> The strictly muscovite language of the Russian revolutionary publications, their lack of attention for the nationalities of Russian Empire other than the Russian one … compel me to oppose the ideas of Russian revolutionaries through my own criticism, which is that of a socialist-federalist Ukrainian.[27]

This Ukrainian socialist federalism was explained in an internationalist journal then edited by Reclus and Perron together with a group of Russian and Slavic exiles, where Dragomanov analysed the oppression of Ukrainian peasants and proposed his idea of a free federation where local and national identities could co-exist in a cosmopolitan context.

> Our cosmopolitanism will not target the destruction of nationalities, which would simply mean the submission of conquered nationalities by the conquering ones and the constitution of privileged social classes. Our cosmopolitanism springs from the revolt of popular classes, from which it will involve all different nationalities – produced by nature – into an international free and egalitarian federation, based on individual autonomy and on the federation of free Communes.[28]

The Proudhonian inspiration of these statements is confirmed in other Dragomanov writings:

> Mankind's aim, which is completely unlike present-day States, is a condition where both larger and smaller social bodies will be composed of free men, united voluntarily for common work and mutual help. This goal is called anarchy. […] Proudhon's anarchism is the doctrine of the complete independence of the individual and the inviolability of his rights by all governmental powers, even elected and representative ones.[29]

Dragomanov's situated cosmopolitanism appeared clearly in his correspondences with the French linguist Gaston Paris, to whom he wrote that his works on the 'folklore' of Eastern Europe had the task of 'building bridges between East and West'.[30]

This simultaneous mobilization of anarchism, national independence and cosmopolitanism was not strange at that time: on the contrary, it characterised an entire generation of militants, including Reclus, who appointed Dragomanov for collaborating on the fifth volume of the *New Universal Geography* (1880) and basically to the long chapter on 'European Russia' (642 pages), which included then Finland, Poland and a great part of Eastern Europe. According to Reclus's biographer Max Nettlau, this corresponded to a federalist political choice, because 'Dragomanov, in his quality of Ukrainian federalist and autonomist, was the better guarantee, for Reclus, to keep his work far from any Russian centralist tendency'.[31] Thus, the *New Universal Geography* militated explicitly for national liberation in Eastern Europe and for the justification of a federalist proposal on geographical bases. According to the two geographers (the volume was signed by Reclus alone, but Dragomanov's participation was largely acknowledged by the French geographer), in the Russian empire 'ethnographic regions do not correspond to the limits of hydrographic basins and openly clash with the administrative provinces' borders, which were often established randomly, or with the clear aim of hindering national affinities'.[32] Thus, ethnography and physical geography were both mobilised against state and administrative borders with the explicit political aim of challenging their situation at the moment.

Reclus's and Dragomanov's work can also be read in the frame of the classical studies on Euro-Orientalism which, according to Larry Wolff, was then the result of a sense of superiority that French Enlightenment felt towards the 'oriental despotism', contrasting with the alleged 'western freedom': 'The Enlightenment had to invent Western Europe and Eastern Europe together, as complementary concepts, defining each other by opposition and adjacency'.[33] Ezequiel Adamovsky observed that, at the end of the nineteenth century, Reclus's *New Universal Geography* was influential in modifying these views starting from basic geographical definitions: 'The most authoritative French academic description of world geography ... reinforced the concept of Eastern Europe and ruled out the location of Russia in the "North", which from then on was meant to be considered "ordinary" (that is, nonscientific) knowledge'.[34] I would argue that the challenge a French anarchist geographer like Reclus addressed to Euro-Orientalism was his willingness to give voice to the Other, by involving a number of scholars from Eastern Europe like Dragomanov, Kropotkin and Metchnikoff in the redaction of his work[35] and presenting Eastern Europe as the place of some of the most advanced revolutionary movements. Reclus argued that the Russian Empire was then the region 'where we find the most outdated forms of absolute power ... and where the innovators launch the most audacious theories for social and political reconstruction'.[36]

Kropotkin, Finland and the geographical invention of nation

If Kropotkin was a classic example of anarchist transnational militant, the rootedness of his ideas has been recently pointed out by several works highlighting the importance of Russian culture for the construction of concepts such as anarchist communism and mutual aid,[37] as well as his insertion in a British intellectual tradition.[38] In the Age of Empire, Finland was an example of nationality striving for its independence and international recognition. As a classical literature on national imagination and invention of tradition showed,[39] a part of these efforts lay in the construction of national histories, cultures, identities, and clearly geographies,[40] by the respective intellectual elites. According to Anne Buttimer, Kropotkin was a reference for the Estonian geographer Edgar Kant (1902–1978), who invented the 'Balto-Skandia' region to justify the independence of Finland and of the 'Baltic Republics' during the interwar period, and who quoted the Russian anarchist geographer as the inspirer of his concept of *Heimatkunde* (knowledge of one's fatherland) mobilising both 'ecological and social dimensions'[41] for this task.

Kropotkin knew Finland very well because he had travelled there when he served the Russian government as an explorer and surveyor. According to his autobiographic memoirs, it was there that he felt definitively uncomfortable with his belonging to an élite (he was born Prince of Smolensk).

> When I was crossing in a Finnish two-wheeled *karria* some plain which offered no interest to the geologist, or when I was walking, hammer on shoulder, from one gravel-pit to another, I could think; and amidst the undoubtedly interesting geological work I was carrying on, one idea, which appealed far more strongly to my inner self than geology persistently worked in my mind. I saw what an immense amount of labour the Finnish peasant spends in clearing the land and in breaking up the hard boulder-clay, and I said to myself: 'I will write the physical geography of this part of Russia, and tell the peasant the best means of cultivating this soil. Here an American stump extractor would be invaluable; there certain methods of manuring would be indicated by science... But what is the use of talking to this peasant about American machines, when he has barely enough bread to live upon from one crop to the next; when the rent which he has to pay for that boulder-clay grows heavier and heavier in proportion to his success in improving the soil? He gnaws at his hard-as-a-stone rye-flour cake which he bakes twice a year; he has with it a morsel of fearfully salted cod and a drink of skimmed milk. How dare I talk to him of American machines, when all that he can raise must be sold to pay rent and taxes? He needs me to live with him, to help him to become the owner or the free occupier of that land. Then he will read books with profit, but not now. ... Science is an excellent thing. I knew its joys and valued them, – perhaps more than many of my colleagues did. Even now, as I was looking on the lakes and the hillocks of Finland, new and beautiful generalizations arose before my eyes ... But what right had I to these highest joys, when all around me was nothing but misery and struggle for a mouldy bit of bread; when whatsoever I should spend to enable me to live in that world of higher emotions must needs be taken from the very mouths of those who grew the wheat and had not bread enough for their children?[42]

Thus, he decided to resign from the Imperial Geographical Society, and to put his knowledge at the service of people's emancipation.[43]

Kropotkin's paper on Finland as a nation was published in 1885 for the popular British journal *The Nineteenth Century*. The geographer's correspondences reveal that this was part of a wider project for writing a book on Finland, which was finally not realised, supposedly due to the difficult material circumstances in which Kropotkin worked in these years. In fact, he wrote this paper when he was a prisoner in France, from December 1882 to January 1886. As shown by recent studies[44] this jail period was very prolific for Kropotkin in scientific terms, because the Russian exile continued, in his cell, to work for Reclus's geography and for the periodical publications of his British editors John Scott Keltie (*Nature* and *Statesman's Year Book*), James Knowles (*The Nineteenth Century*) and William Roberson-Smith (*Encyclopaedia Britannica*). Kropotkin's jail correspondences with them witness the width of the original project on Finland and the Reclus's role in inspiring it. As the prisoner wrote to Scott Keltie in 1883:

> In a few days I shall have a separate room to work therein, and I hope to undertake some larger work, namely a complete description of Finland, a young rising state. Some parts of it shall be written in such way as to be published in English reviews, and the whole would make a book. Reclus strongly recommends to me to make a book on Finland.[45]

From his 'centre of calculation' in Switzerland, Reclus contributed to Kropotkin's work sending him books, papers and study materials on Finland.

> I send you the list of the papers concerning Finland which have been published in the Petermann's *Mitteilungen*. About the complete bibliography, we can have it in Paris, because the Hachette library has all the collection. Otherwise, we can ask the Geographical Society. Moreover, we can have through the librarian, M. James Jackson, the work by Valfrid Vasenius, *Suomalainen Kirjallisuus 1544–1699*, with the alphabetic catalogue on existing works on Finland.[46]

Reclus's interest is clearly due to political reasons of the same kind as his collaboration with Dragomanov, i.e. the strategic aim of challenging autocracy by supporting rising nationalities; this is even clearer if we consider that the national monographs, at that time, were instruments for nation building. Thus, nineteenth-century anarchist geographies aimed to play an active and performative role in the wider movement of the geographical invention of nations.

Finally, Kropotkin managed to publish a long paper for the *Nineteenth Century* because its editor, Knowles, was interested in denouncing the treatment of political prisoners in the Russian Empire before international public opinion. He wrote to Kropotkin, in a rather astonishing way if we consider that the geographer was then a political prisoner in France, that the 'Russian treatment of political prisoners [had] to be brought to the knowledge of the highest authorities'.[47] Kropotkin's paper established a link between the Finnish cause of that day and the revolutionary republican tradition of 1848. 'National questions are not in vogue now in Europe. After having so much excited the generation of 1848, they seem to be now in neglect [for the] poor results of a movement which caused so many illusions.'[48] The reason for this disillusion, according to Kropotkin, was that the national revolution had failed in realising contextually the social one.

> Italian unity has not improved the lower classes of the Peninsula, and they have now to bear the burden of a State endeavouring to conquer a place among the great Powers. The formerly oppressed Hungary is oppressing in her turn the Slavonic populations under her rule. The last Polish insurrection was crushed rather by the agrarian measures of the Russian Government than by its armies and scaffold.[49]

Nevertheless, according to Kropotkin, the national question existed and deserved to be considered by revolutionaries. 'Notwithstanding all this, national questions are as real in Europe as ever, and it would be as unwise to shut our eyes to them as to deny their importance.'[50]

A European panorama of ongoing national claims reveals Kropotkin's sympathy for them, and his proposal for federalist solutions, which recalls at the same time Reclus's and Dragomanov's statements and classical anarchist themes by Proudhon and Bakunin.

> Irish Home Rule, the Schleswig 'difficulty,' and Norwegian 'separatism' are problems which must be resolved ... Not only a thorough discontent, but a chronic insurrectionary agitation is going on among the Serbo-Croats, who are endeavouring to shake off the yoke of Hungary. The Czechs, the Slovaks, the Poles of Austria are struggling, too, for self-government; as also, to some extent, the Slowens, or Wends, and the Little Russians of Eastern Galicia; while neither peace nor regular development is possible on the Balkan Peninsula until the Bosnians, the Herzegovinians, the Serbs, the Bulgarians, and others, have freed themselves from Turkish rule, Russian 'protection', and Austrian 'occupation', and have succeeded in constituting a free South-Slavonian Federation ... Finally, in the North-east we have Finland, where, one of the most interesting autonomist movements of our time has been steadily going on for more than sixty years.[51]

If one considers the aforementioned idea of rooted transnationalism, it is not surprising that a 'Founding Father' of anarchism like Kropotkin endorsed the love people feel for their country; in the same vein, it is clear that these systematic international comparisons showed that the problem was not that of one nationality, but a common issue for every people. 'Everybody loves his own country: with the Finns this love becomes a passion, as powerful as the passion of the Scottish Highlander for his 'land of mountain and of flood'; and it has the same source'.[52] For Kropotkin, the national question was also an analytical tool for addressing the hierarchies of oppression, which he considered as complex and multi-fold. Social oppression toward the lowest classes was accompanied by a national hierarchy, whose most prestigious exponents were the members of Russian administration and Swedish old aristocracy.

> It is not so with the Swedish nobility, Swedish tradesmen and Swedish officials, until now they have constituted the dominant element in Finland's political and economic life; they are still landholders in a larger proportion than the Finns; and, by maintaining Swedish as the official language in the Administration, they have systematically eliminated from it the Finnish element, which they still regard with contempt … Thus the struggle is not one between two races, it is for the maintenance of class privileges inherited from the Swedish domination.[53]

In Kropotkin's historical analyses, the ancient Swedish domination had at least the effect of accustoming Finnish people to different juridical institutions than those supported by Russian peasants, and this sharpened the local spirit of revolt against the new masters.

> Finland now cherishes the hope of becoming an independent State herself [also because] Swedish rule again saved Finland from serfdom – at least from the disgrace of personal servitude, and it accustomed the peasant to the sound of his own voice in the State's representation.[54]

The activism of local patriots was then praised by Kropotkin, who argued that: 'It was necessary to prove to the indifferent that the watchword, "Finland for the Finns", is not an empty dream, but may become yet a reality',[55] considering the Finnish movement as more radical than other national movements then challenging the Czar. '[Finland's] national movement does not ask a return to the past, as has been the case with Poland; it aspires after a quite new, autonomous Finland.'[56]

The Russian geographer paid attention to attempts for revitalising Finnish language, cultures and traditions in order to spread national consciousness, and called for an extension of popular education in this sense. 'It is obvious that the more national consciousness is raised in Finland, and the more education is spread among its people, the more will it feel the weight of Russian sovereignty.'[57] Kropotkin's conclusions highlighted the geopolitical importance of nationalities in 'redrawing the map of Europe' and endorsed national claims as steps toward a more radical social revolution:

> Only one thing is certain: that the ardour of Finnish patriots for awakening among their people national feeling and the longing for a complete

independence will be redoubled by the attempts, recently renewed, against Finland's autonomy. The map of Europe has already undergone many changes, and it is not improbable that the social and political complications which accumulate on Old Europe's head may result, among other things, in the restoration of Finland to the Finns.[58]

'The laboratory for miscegenation': Reclus and Brazil

During his long career of geographer and activist, Elisée Reclus analysed a great number of nationalities all over the world. The example of Brazil is very fit to address his approach to the idea of nation because it is linked to the topics of both de-colonisation and federalism. Reclus's first writings on Brazil, published in the 1860s for the French journal *La Revue des Deux Mondes*, harshly criticised Brazilian élites for the centralism of their Empire and the persistency of slavery and feudal institutions.[59] In a series of papers on the 'War of Paraguay', published in the same journal from 1864 to 1868, Reclus criticised the alliance between Argentina, Uruguay and Brazil against Paraguay, because he considered the mutual hostility among recently de-colonised republics as something like a fratricide combat. According to Reclus, the common enemy should have been the Empire of Brazil, which the geographer compared explicitly to the French Second Empire, using this overseas war as an indirect way to intervene in the French debate, when he was taking part in the most radical republican opposition to Napoleon III.[60] These results are clear if one considers his use of expressions like 'a centralising code like that of Brazil or Imperial France'.[61] In Reclus's analysis, the true war was 'between the slave-owning oligarchy and republican democracy',[62] considering that Brazil was then 'an empire when slavery reigns, where territory is owned by a handful of great proprietors and where women are subject to a sort of reclusion'.[63] In these years, Reclus showed an interest for the federalism of some Latin-American republics and seemed to be particularly interested in the autonomist movements of the Northern Argentinian regions Entre Rios and Corrientes; he also corresponded on these topics with the Argentinian republican Juan Bautista Alberdi.[64]

Between 1888 and 1889, the abolition of slavery and the transformation of Brazil in a federalist republic led Reclus to change completely his mind on the possible geopolitical role of the country. According to the French geographer, the abolition of slavery was such a powerful reform that it also explained the collapse of imperial institutions: 'At the same time, the political form of Brazil was modified; the centralist empire became suddenly a federal republic'.[65] The new republic was seen by Reclus as one of the laboratories for his programme of global miscegenation intended as a way for definitively getting rid of racism: 'Brazil is the promised land where humankind, represented by Whites, Reds and Blacks, is brotherly merging and mutually reconciling.'[66] Reclus praised the measures taken by Brazilian government to favour migration by 'declaring Brazilians all the residents of foreign origins ... who did not claim another nationality'.[67] It is worth noting that, at that time, proletarians' migration toward Latin America, and especially in Brazil, Uruguay and Argentina, was bringing their socialist and anarchist ideas,[68] thus Reclus's favour for this global movement is clear to understand. Nevertheless, the anarchist geographer was not only an apologist of the new republic: he often highlighted its contradictions, like the persistence of latifundia owned by the descendants of ancient Portuguese colonisers. 'The Republic ignores aristocratic privileges ... and abolishes noble titles, but in few other countries one finds so many counts and marquises'.[69] The democratic process, then, was biased by the fact that the 'federalist republic has been proclaimed but ... people have not been consulted'.[70]

In 1893, Reclus had the opportunity to visit Brazil, staying a few months in the country. During his visit in Rio de Janeiro, the geographer was received with full honours by the most important cultural institutions of the new Republic, namely the Geographical Society, the Brazilian Institute of History and Geography and the Brazilian Academy of Letters.[71] One of the reasons for this triumphal reception and for Reclus's later popularity in Brazil, was his acquaintance with José María da Silva Paranhos Júnior (1845–1912), known as Baron Rio Branco and considered one of the Founding Fathers of modern Brazil. Rio Branco, also nicknamed 'the Brazilian Bismarck', lived then in Europe, where he performed hard diplomatic work on the numerous border disputes Brazil had, in these years, with Argentina, Bolivia and French Guiana, which were always resolved in favour of Brazil, also thanks to Rio Branco's activism and cleverness.[72] In April 1893, Rio Branco asked Reclus a geographical expertise on the contested area between Brazil and Argentina,[73] giving him in exchange geographical information for the *New Universal Geography*.[74] At this stage of my research, I don't know exactly the terms of Reclus's intervention in this border dispute, but there is evidence that the anarchist geographer had contacts with scholars of both nationalities, as this dispute was evoked, some years later, by an anonymous correspondent from Buenos Aires.[75]

In 1900, the chapter dedicated to Brazil in the *New Universal Geography* was edited and translated into Portuguese by Rio Branco and Ramiz Galvão under the title *Estados Unidos do Brasil*.[76] This publication had a great importance for the geographical definition of Brazil as a nation, because it was the first national monograph of the country, and its regionalisation was widely used by Brazilian geographers of the following generation like Delgado de Carvalho.[77] The importance of this translation as part of the work of nation-building is confirmed by the critical apparatus (substantial introduction and endnotes) provided by the Brazilian editors, who 'adapted' Reclus's geography to their geopolitical concerns, i.e. the consolidation of Brazilian influence in Latin America, as recent Brazilian research shows.[78] What is ironic is that Rio Branco, in his introduction, criticized Reclus about the issue of the Franco-Brazilian border dispute, considering Reclus's position not enough filo-Brazilian, as the French geographer, in his work, simply claimed neutrality on a border controversy which he deemed absurd, arguing that 'one could fill the libraries with reports and diplomatic documents issued on this unsolvable question'.[79] But a few months later, in December 1900, the diplomatic referee chosen by the two disputants, i.e. the government of the Helvetic Confederation, attributed all the contested territory to Brazil,[80] including geographical arguments taken from Reclus's geography in the official motivations, e.g. his remarks on the interior disputed region among the coast of the present-day state of Amapa and the valley of the Rio Branco River. According to Reclus, this formally disputed territory, of almost 300,000 square kilometres, had already 'become uncontestably Brazilian by language, customs, and political and commercial relations'.[81]

In subsequent years, Reclus's influence in Brazil reached both progressive liberal intellectuals and socialist workers: in the first case this occurred thanks to his geographical works, both in French and in the Galvão translation; in the second case it was thanks to his pamphlets of anarchist propaganda mainly translated and printed in Portugal.[82] After 1900, a new frontier dispute occupied the Brazilian diplomacy: this time it was for the definition of the Southern borders of British Guiana. The idea of asking again the help of the elderly anarchist geographer arose and a diplomatic expedition joined Reclus in Brussels, in 1903. A member of the Brazilian group was an old Reclus admirer, José Pereira de Graça Aranha (1868–1931), a famous writer and member of the progressive Brazilian bourgeoisie during the first years of the Brazilian Republic. According to Aranha's biographer Maria Helena Castro Azevedo, the Brazilian intellectual never forgot his youthful sympathy for anarchism and continued to read and admire Kropotkin and Reclus. Castro Azevedo reported that, in Brussels, 'Reclus accepted the task in order to oppose the interests of the British in the passage to the Amazon River, where they intended to build a railway; he was also personally very interested in this region, as he wanted America for Americans'.[83]

According to Castro Azevedo, this implied Reclus's tactic adhesion to the 'Monroe Doctrine'. I would argue that Reclus, who considered this doctrine as one 'of which it is worth distrusting',[84] did not prefer Brazilian imperialism to the British one, but that his empathy towards de-colonised South American republics belonged to a time and a militant culture where a republican nation was seen as a step towards more ambitious revolutionary outcomes. This position had clear links with the early anti-colonialism of anarchist militants in the Age of Empire, which concerned at the same time the extra-European world[85] and the colonial relations within Europe, as shown by the collaboration between Dragomanov and Reclus.[86] In his writings, Reclus called explicitly for the anticolonial revolt of all colonised peoples (it is worth stressing that, at that time, the word 'nation' could be used as synonymous for 'people'):

How could it be otherwise? This hatred of the slave who revolts against us is right, and proves at least that there is still hope of emancipation. It is natural that the Hindus, Egyptians, Kaffirs and Irishmen hate Englishmen; it is natural that Arabians execrate Europeans. That's justice![87]

Thus, Reclus's relation with the Brazilian nation concerned both his geographical published works and his political and social networks; his geography was not limited to the study of Brazil as a geographical object, but it had an agency in local politics and diplomacy: later it would become a reference for local socialist and anarchist movements and especially for the experiences of anarchist education performed by the 'Modern Schools'.[88]

In any case, the Brazilian edition of the *New Universal Geography*'s chapter on Brazil by Rio Branco and Ramiz Galvão, assumed the role of national monograph on Brazil like the geographical monographs published in Europe in the same years, which become foundational works for the geographical invention of nation and the formation of the respective national geographic schools. It was the case, among others, of the *Tableau de la géographie de la France* by Paul Vidal de la Blache (1903), *Britain and British Seas* by Halford Mackinder (1902), *Deutschland: Einführung in die Heimatkunde* by Friedrich Ratzel (1899).[89] Translations of Reclus's chapters often played this role in Latin America, like in the case of Colombia,[90] and in Southern Europe, like in the case of Italy,[91] but a systematic research on these translations of Reclus's geography has still to be done and it could undoubtedly give a further contribution to the study of the relations between anarchism and the idea of nation in the Age of Empire.

Conclusion: anarchism, nationalism and cosmopolitanism

This chapter has shown that the early relation between anarchism and struggles for national liberation led this movement to a pioneering critique of both internal and external colonialisms. This was not in contradiction with the transnational nature of the anarchist movement because the cases of Reclus, Dragomanov and Kropotkin showed the importance of the exile networks and of the international circulation of knowledge in the construction of what anarchists considered as an international solidarity in favour of the 'oppressed' nations: Reclus and Kropotkin, for instance, did not call for their own nations as in the case of the most classical nationalist anti-colonial movements, but worked against the colonial interests of the respective 'homelands', respectively France and Russia.

Moreover, as Dragomanov showed during his Geneva years, forms of situated belonging are not incompatible with internationalism and cosmopolitanism. If places and circulations play a role in the construction of knowledge, this cosmopolitanism was based on the transnational nature of the anarchist movement, recently studied by historians, and it is not a coincidence that we find geographers, accustomed to dealing with spaces and places, among the militants well aware of the importance of local identities and embodiments.

Thus, historical anarchist geographers believed in internationalism without neglecting local identities and cultural differences as instruments for a critique of the empires, being aware, at least empirically, of the importance of situating struggles. Further research is needed on their approaches to stateless peoples like Jews, Armenians and Gypsies, which they nevertheless acknowledged as nations, raging at the persecutions they suffered. Furthermore, the genealogical and situated links between anarchism and republicanism still need further exploration.

Notes

1 See for this concept: Hobsbawm, *The Age of Empire*.
2 Ferretti, *Élisée Reclus: pour une géographie nouvelle*; Pelletier, *Géographie et anarchie: Reclus, Kropotkine, Metchnikoff*.
3 Vuilleumier, *Histoire et combats. Mouvement ouvrier et socialisme en Suisse, 1864–1960*.
4 Springer et al., 'Reanimating Anarchist Geographies: A New Burst of Colour'; Springer, *The Anarchist Roots of Geography: Toward Spatial Emancipation*.
5 Bantman, *The French Anarchists in London, 1880–1914. Exile and Transnationalism in the First Globalization*; Bantman and Altena, *Reassessing the Transnational Turn: Scales of Analysis in Anarchist and Syndicalist Studies*; Di Paola, *The Knights Errant of Anarchy, London and the Italian Anarchist Diaspora (1880–1917)*; Hirsch and van der Walt, *Anarchism and Syndicalism in the Colonial and Postcolonial World, 1870–1940*; Turcato, 'Italian Anarchism as a Transnational Movement, 1885–1915'.
6 Graham, *We Do not Fear Anarchy, We Invoke It! The First International and the Origins of the Anarchist Movement*; Marshall, *Demanding the Impossible: A History of Anarchism*.
7 Turcato, 'Nations without Borders: Anarchists and National Identity', in Bantman and Altena (eds), *Reassessing the Transnational Turn: Scales of Analysis in Anarchist and Syndicalist Studies*, 30.
8 Ferretti and Castleton, 'Fédéralisme, identités nationales et critique des frontières naturelles: Pierre-Joseph Proudhon (1809–1865) géographe des "États-Unis d'Europe"'; Hayat, 'Le fédéralisme proudhonien à l'épreuve des nationalités', in J. Cagiao y Conde (ed.), *Le fédéralisme: le retour?* '.
9 Levy, 'Anarchism, Internationalism and Nationalism in Europe, 1860–1939'.
10 Nettlau, *La Première Internationale en Espagne (1868–1888)*.
11 Turcato, *Making Sense of Anarchism: Errico Malatesta's Experiments with Revolution, 1889–1900*, 17.
12 Masini, *Eresie dell'Ottocento: Alle sorgenti laiche, umaniste e libertarie della democrazia italiana*.
13 Lehning, 'Michel Bakounine et le Risorgimento tradito'.
14 Nettlau, *La Première Internationale*, 33.
15 Ferretti, *Élisée Reclus*.
16 Manfredonia, *Les anarchistes et la Révolution française*.
17 Bakunin, *Oeuvres*, 254.
18 *Anarchism and Syndicalism in the Colonial and Postcolonial World, 1870–1940*.
19 *Under Three Flags: Anarchism and the Anti-Colonial Imagination*.
20 Ferretti, '"They Have the Right to Throw Us Out": Élisée Reclus' *New Universal Geography*'; 'Arcangelo Ghisleri and the "Right to Barbarity": Geography and Anti-Colonialism in Italy in the Age of Empire (1875–1914)'.
21 Ramnath, *Decolonizing Anarchism: An Antiauthoritarian History of India's Liberation Struggle*.

22 Konishi, *Anarchist Modernity: Cooperatism and Japanese-Russian Intellectual Relations in Modern Japan*, 3–4.
23 Gramsci, *The Modern Prince and other Writings*; Hechter, *Internal Colonialism: The Celtic Fringe in British National Development, 1536–1966*.
24 Pettit, *Republicanism: A Theory of Freedom and Government*.
25 *Correspondance de Michel Bakounine, Lettres à Herzen et à Ogareff (1860–1874), Publiées, avec Préface et Annotation par Michel Dragomanov.*
26 Bankowski-Züllig, '"Kleinrussischer Separatist oder ‚Russischer Nihilist?" M.P. Drahomanov und die Anfange der ukrainischen Emigration in der Schweiz', in Bankowski-Zullig, Urech and Brang (eds), *Asyl und Aufenthalt: die Schweiz als Zuflucht und Wirkungsstätte von Slaven im 19. und 20. Jahrhundert*; Mysyrowicz, 'Agents secrets tsaristes et révolutionnaires russes à Genève 1879–1903'.
27 Dragomanov, *Le tyrannicide en Russie et l'œuvre de l'Europe occidentale*, 3.
28 Dragomanov, 'Les paysans Russo-ukrainiens sous les libéraux Hongrois', 14.
29 Dragomanov, *A Symposium and Selected Writings*, 73–74.
30 Bibliothèque Nationale de France, Nouvelles Acquisitions Françaises, 24438, f. 169, M. Dragomanov to G. Paris, 8 June 1880.
31 Nettlau, *Eliseo Reclus: vida de un sabio justo y rebelde*, 79.
32 Reclus, *Nouvelle Géographie Universelle*[hereafter *NGU*], vol. V, 487.
33 Wolff, *Inventing Eastern Europe: The Map of Civilization in the Mind of the Enlightenment*, 165.
34 Adamovsky, 'Euro-Orientalism', 607.
35 Nettlau, *Eliseo Reclus*, 79.
36 Reclus, *NGU*, vol. V, 892.
37 Livingstone, 'Science, Text and Space: Thoughts on the Geography of Reading'; Kinna, *Kropotkin: Reviewing the Classical Anarchist Tradition*.
38 Adams, *Kropotkin, Read and the Intellectual History of British Anarchism*; Kearns, 'The Political Pivot of Geography'; McLaughlin, *Kropotkin and the Anarchist Intellectual Tradition*.
39 Anderson, *Imagined Communities: Reflections on the Origin and Spread of Nationalism*; Hobsbawm, *The Invention of Tradition*.
40 Ferretti, 'Inventing Italy. Geography, Risorgimento and National Imagination: the International Circulation of Geographical Knowledge in the 19th Century'; Hooson, *Geography and National Identity*.
41 Buttimer, 'Edgar Kant and *Heimatkunde*: Balto-Skandia and Regional Identity', in Hooson (ed.), *Geography and National Identity*', 166.
42 Kropotkin, *Memoirs of a Revolutionist*, 18, 21.
43 Woodcock and Avakumović, *The Anarchist Prince: A Biographical Study of Peter Kropotkin*.
44 Ferretti, 'The Correspondence between Élisée Reclus and Pëtr Kropotkin as a Source for the History of Geography'.
45 Royal Geographical Society–Institute of British Geographers, Manuscripts Department CB7, Kropotkin to John Scott Keltie, 3 April 1883.
46 Gosudarstvennyi Arkhiv Rossiiskoi Federatsii [hereafter GARF], Fondy P-1129, op. 2 khr, 2103, E. Reclus to P. Kropotkin, 10 April 1883.
47 GARF, 1129, 2, 1895, Knowles to Kropotkin, 15 November 1884.
48 Kropotkin, 'Finland: A Rising Nationality', 1.
49 Kropotkin, 'Finland', 1.
50 Kropotkin, 'Finland', 1.
51 Kropotkin, 'Finland', 1–2.
52 Kropotkin, 'Finland', 4.
53 Kropotkin, 'Finland', 4–5.
54 Kropotkin, 'Finland', 6.
55 Kropotkin, 'Finland', 8.

56 Kropotkin, 'Finland', 12.
57 Kropotkin, 'Finland', 14.
58 Kropotkin, 'Finland', 14–15.
59 Reclus, 'Le Brésil et la colonisation, I. Le bassin des Amazones et les Indiens'; 'Le Brésil et la colonisation, II. Les provinces du littoral, les noirs et les colonies allemandes'.
60 Ferretti, 'Républicanisme'.
61 Reclus, 'L'élection présidentielle de La Plata et la guerre du Paraguay', 899.
62 Reclus, 'La guerre de l'Uruguay et les républiques de La Plata', 968.
63 Reclus, 'La guerre de l'Uruguay', 968.
64 Quattrocchi-Woisson, *Juan Bautista Alberdi y la Independencia Argentina. La fuerza del pensamiento y de la escritura.*
65 Reclus, *NGU*, vol. XIX, 104.
66 Reclus, *NGU*, vol. XIX, 112–113.
67 Reclus, *NGU*, vol. XIX, 487.
68 Shaffer, 'Latin Lines and Dots: Transnational Anarchism, Regional Networks, and Italian Libertarians in Latin America'.
69 Reclus, *NGU*, vol. XIX, 486.
70 Reclus, *NGU*, vol. XIX, 488.
71 Carris Cardoso, *O lugar da geografia brasileira: a Sociedade de Geografia do Rio de Janeiro entre 1883 e 1945*; Miyahiro, *O Brasil de Elisée Reclus: territorio e sociedade em fins do século XIX.*
72 Romani, *Aqui começ o Brasil! Histórias das gentes e dos poderes na fronteira do Oiapoque.*
73 Rio de Janeiro, Itamaraty Archives, Archive Rio Branco, Rio Branco to Reclus, 10 June 1893. Thanks to Luciene Carris Cardoso and David Ramirez Palacios for sharing with me these materials.
74 Itamaraty Archives, Archive Rio Branco, Reclus to Rio Branco, 6 April 1893.
75 Bibliothèque Nationale de France, Nouvelles Acquisitions Françaises, dossier 22914.
76 Reclus, *Estados Unidos do Brasil, Geographia, Ethnographia, Estatistica.*
77 Delgado de Carvalho, *Metodologia do Ensino Geográfico.*
78 Miyahiro, *O Brasil de Elisée Reclus.*
79 Reclus *NGU*, vol. XIX, 86.
80 Ferretti, 'A New Map of the Franco-Brazilian Border Dispute (1900)'.
81 Reclus *NGU*, vol. XIX, 85.
82 Lopes, 'Élisée Reclus e o Brasil'.
83 Castro Azevedo, *Um senhor modernista: biografia de Graça Aranha*, 77.
84 Reclus, *L'Homme et la Terre*, 125.
85 Anderson, *Under Three Flags: Anarchism and the Anti-Colonial Imagination*; Hirsch and van der Walt, *Anarchism and Syndicalism.*
86 Ferretti, 'They Have the Right to Throw Us Out'.
87 Reclus, 'Léopold de Saussurre, Psychologie de la colonisation française dans ses rapports avec les sociétés indigènes'.
88 Romani, 'Da Biblioteca Popular à Escola Moderna. Breve história da ciência e educação libertária na América do Sul'.
89 Capel, *Filosofia y ciencia en la geografia contemporanea*; Robic, *Le tableau de la géographie de la France de Paul Vidal de la Blache: dans le Labyrinthe des Formes.*
90 Ramirez Palacios, *Elisée Reclus e a geografia da Colômbia: cartografia de uma interseção.*
91 Ferretti, 'Traduire Reclus: l'Italie écrite par Attilio Brunialti'.

7 Historicising 'anarchist geography'

Six issues for debate from a historian's point of view[1]

Pascale Siegrist

The past decade has witnessed an important upsurge of anarchist approaches in geography – this volume is itself testimony to this development. The turn to anarchist theory, inspired also by recent grassroots movements, constitutes a rupture within the field of radical geography insofar as it questions the Marxist approaches that had been instrumental for the establishment of radical geography at universities since the 1970s. As Simon Springer, arguably the most polemical of current anarchist geographers, strives to give the deathblow to the already waning influence of Marxism,[2] he among others deploys a plainly historical type of argument: anarchism's connection to geography largely predates that of Marxism. The roots of radical geography, he holds, stretch back into the nineteenth century – a view that even Springer's main target, the Marxist geographer David Harvey has never challenged. The latter had long before conceded that 'the radical urge in nineteenth-century geography was expressed through anarchism rather than through Marxism'.[3] For Marxists and anarchists alike, the inevitable points of reference are Élisée Reclus (1830–1905) and Pëtr Kropotkin (1842–1921). The Frenchman and the Russian, who knew each other and at times collaborated,[4] in their days were widely known as both anarchist militants and acclaimed geographers. The present surge of interest in anarchism is thus framed as a *re*discovery of a forgotten – if not deliberately ousted – tradition.[5] Whilst Reclus has already served as a figurehead for the kind of more consciously political (and left-wing) geography promoted by the journal *Hérodote* since the 1970s[6] and early issues of *Antipode* had been devoted to Kropotkin,[7] Springer wants to take the appropriation yet further. With reference to 'radix' as the root of the term radicalism, Springer poses the question: 'Yet how could a "radical" geography truly be radical without digging down into the foundations that had been laid by the anarchist geographies of Elisée Reclus and Peter Kropotkin?'[8] The argument runs that a thorough engagement with history is now mandatory for radical geographers.

Simultaneously, and sometimes in direct exchange with Anglophone colleagues, geographers in France and Italy have 'dug deeper' into these foundations. Historically minded geographers like Philippe Pelletier and Federico Ferretti have re-examined the writings of Reclus from the point of view of both politics and geography.[9] Unearthing new biographical material from archives across Europe, Ferretti in particular has contributed to alter our perception of Reclus and Kropotkin as eccentric figures at the margins of the field – showing how the anarchists were well respected and in touch with contemporary geography. These studies, moreover, have brought to light a much larger transnational network comprising a good dozen individuals with anarchistic and geographical leanings: a list of nineteenth-century founding fathers of anarchist geography would now include Reclus' closest collaborators Lev Il'ich Mechnikov, Charles Perron, Mikhailo Drahomanov, Gustave Lefrançais, Attila de Gerando as well as further members of his family – his brothers Élie and Onésime Reclus and his cousin Franz Schrader. Better-known figures in the history of the left-wing engagement with space, most notably Patrick Geddes (who never identified as anarchist), have been brought closer to anarchism by virtue of their proximity to Reclus and Kropotkin. Given this remarkable personal overlap, the shortcut from anarchist geographers to anarchist geography seems but a small step. Like Yves Lacoste and Béatrice Giblin in the decades before them,[10] Ferretti and Pelletier are both interested in exploring anarchism as an epistemological point of departure for geography. Despite their overwhelmingly historical perspective, they regard this early coincidence of anarchy and geography as a fertile source of inspiration for today's geographies.[11] In this ambition, their programme smoothly complements that of Springer et al.[12]

However, pinpointing where exactly the 'anarchist' element in Reclus and Kropotkin's geography resides has proven exceedingly difficult. Pelletier forestalls that 'the relations between geography and politics are not always very clear, nor explicit'.[13] Anarchy and geography surely coalesce in specific issues (the state, frontiers, the unity of mankind, respect for nature, rejection of imperial conquest, the stress on education, etc.), but a geography informed by anarchist standpoints is hardly the same as one that is premised on 'anarchism' epistemically. Matters are further complicated by the fact that there are noticeable differences in the approaches pursued by the geographer-anarchists of the period. Ferretti notes that the work produced by geographers associated with Reclus and his publisher Hachette hardly forms a coherent methodological whole – 'because of its heterogeneity, it had perhaps not yet arrived at constituting a kind of scientific paradigm in the Kuhnian sense'.[14] This raises the question of – to stay within Kuhnian terminology – whether nineteenth-century anarchist science might have been a lot more 'normal' after all. Important parallels to prevailing non-anarchist modes of writing the world would suggest so. The lack of a broader social-theoretical framework rendering geography discernibly radical is also Harvey's main point of criticism.[15] Responding to Springer's argument, Harvey denies that politics played a decisive role in shaping much of the anarchists' geographical work: 'Looking at [Reclus'] nineteen volume *Geographie Universelle*, there is little trace of anarchist sentiments (any more than there were in Kropotkin's studies of the physical geography of central Asia).'[16] Taking the point even further, Harvey is led to conclude that '[h]istorically there has been a separation between geographical work and politics'.[17] At the other end of the debate as well, history is called upon to provide the decisive argument.

It is thus timely for historians to weigh in on the debate. Historians have struggled no less than geographers to capture the double identity of Reclus and Kropotkin as anarchists and geographers: histories of either anarchism or geography tend to deal with the respective 'other' affinity as a curiosity noted in passing.[18] The dichotomy is further amplified by a disproportionate treatment of Kropotkin as anarchist and Reclus as geographer. Reclus' geographical work at all times figures prominently in analyses[19] and he emerges as the centre of gravity of the field of anarchist geography as mapped out by Pelletier and Ferretti. But his position within anarchist historiography compares unfavourably to Kropotkin's.[20] On the latter, we possess a series of specialist monographs,[21] but with the exception of a Soviet study from 1952,[22] none of the historical studies focus squarely on his activities as a geographer or his geographical thought.[23] The first book-length study by a geographer only came out in 2016.[24] Historians tend to present the world of science and that of revolutionary politics as a source of contradiction; for Martin A. Miller, Kropotkin 'belonged to both worlds, the one he was working to destroy and the one which was to replace it in the future'.[25] Historians ought to take seriously the interdependencies of both worlds, the complex entanglement of anarchy and geography that (radical) geographers have long been pointing out. On the other hand, reading Kropotkin and Reclus in their own right and time, with an increased sensitivity for characteristically 'historical' issues like chronology and context, serves as a precaution against blurring contemporary notions of geography with historical ones. This holds the potential of providing us with a more nuanced picture of the geography of radicals and the possibilities and limits of a radical geography in the late nineteenth century.

With this chapter, I therefore hope to enhance our understanding of these crucial historical figures by applying several strategies from the historian's toolbox. Understanding the problem of the relationship between geography and anarchy as both a problem of epistemology and one of political thought, the selected approaches are essentially those of intellectual historians of different schools: the history of concepts, contextualism, genealogy. Thus, while not denying the fundamental importance of the direct exchange between theory and practice for anarchism, this chapter is primarily intended as a contribution to the geographical thought of nineteenth-century anarchists. Its approach consists in historically reconstructing their conceptualisation of geography and its aims, without taking for granted some of the presuppositions of contemporary radical geography. It is not my aim to provide a mere stage stetting, a historical 'backdrop' on which to draw. Rather, I regard (intellectual) history as yielding conclusions in its own right about the nature of the object under scrutiny – a position that, incidentally, is also that of Reclus and Kropotkin, who believed that to understand a situation or an idea, it was important to see how it had come about. I obviously do not mean to make a case for an incompatibility between history and geography, but for limits of space and in view of Springer's invitation to cultivate an 'agonistic' debate, I will focus on what I take to be the potential areas of disagreement. I argue that anarchist geography is largely a modern construct that, as it happens, obliterates much of the way in which Reclus and Kropotkin thought about the relationship between science and radicalism. I structure these broad and interrelated arguments around six main points.

The history of concepts: ἀναρχία and γεωγραφία

For intellectual historians, and for conceptual historians in particular, words do matter. The origin of concepts, their usage, and the changes that occur therein are taken to be indicative of larger shifts in our mentality and culture.[26] The first difficulty when projecting a conception of geography founded on anarchist principles back into the nineteenth century is, quite bluntly, one of terminology. To my knowledge, neither Reclus, nor Kropotkin, nor any of their colleagues and comrades ever speaks of 'anarchist geography'. The association that has become a commonplace in radical geography is so hard to track down to concrete passages that it seems difficult to counter Harvey's statement cited above: in their strictly speaking geographical works, the anarchism is subliminal at best. In writings that can be deemed geographical on grounds of their publication in geographical journals or because of their declaration as such, the concept 'anarchism' is virtually absent – except for the few mentions of 'anarchic' and related terms in Reclus' *L'Homme et la Terre* (where it appears mainly as a trait characteristic for groups of people that he takes to be particularly independent; Bedouins and Arabs, *raskolniks* and other religious dissenters).[27] For a conceptual historian wanting to go back to the origins of 'anarchist geography', this is a serious problem. The authors' own conception of what fell into the category of 'geography' strikes as an incontestable yardstick to begin with. Against the ever-looming danger of anachronism, Gad Prudovsky has made a convincing case that it can indeed be possible in the history of science to 'ascribe to past thinkers concepts they had no linguistic means to express'.[28] Since Kropotkin and Reclus made abundant use of both 'anarchy' and 'geography' as terms in their work, the linguistic means were in place. The political, scientific, and sociological ones might however not have been.

The most obvious explanation for the lack of explicitness in the formulation of an 'anarchist geography' is of course censorship by a scientific community with little sympathy for anarchism. Harvey suggests editorial restrictions as the reason for the feeble presence of anarchist arguments in Reclus' geography and Alison Blunt and Jane Wills regret that 'Kropotkin and Reclus were not able to combine their anarchist ideas with their geographical scholarship as they might do today'.[29] When Reclus proposed a book project to his publisher Hachette in 1872, his brother Onésime (who worked for Hachette) warned him that his 'descriptive geography is not to be politico-religio-sociologico-militant', since the publishers were 'obviously worried about a *livre de combat*.[30] (Self-)censorship might have had a role to play, but Ferretti is right to insist on the high degree of liberty that Reclus eventually managed to secure from his editors.[31] All descriptions of Reclus stress his uncompromising devotion to principles; in continuity with his strict Protestant upbringing, he refused to submit to any other guidance than that of his conscience. Rejecting an offer of early liberation from prison on condition that he signed a formal statement to the Geographical Society, he explained his decision to his wife saying: 'The future being unknown to me, it is absolutely impossible to know which course of action my conscience will command, and consequentially, I cannot subscribe to any engagement whose terms will be deliberated by others than myself.'[32] I doubt that he would have accepted any restrictions imposed on his science, especially not for rewards like fame or money that he attached so little value to. His dream of building a Great Globe, 127 meters in diameter, for the Paris Universal Exhibition of 1900 failed mainly because of his refusal to sacrifice scientific rigour in order to make the project more attractive to potential sponsors.[33] Kropotkin, similarly, is rumoured to have rejected a professorship at Cambridge because it would have compromised his freedom of political expression.[34] Regardless of whether the outcome would have been different in a climate of greater academic liberty, what is decisive is that Reclus and Kropotkin both felt it was possible and not in conflict with their convictions to write geography that was devoid of openly anarchist characteristics.

If editorial provisions are a plausible explanation for the moderation in scientific texts, they cannot account for the same observation regarding Reclus and Kropotkin's contributions to the anarchist press: in Reclus and Kropotkin's anarchist articles *vice versa* the straightforward reference to geography is wanting. Freedom of expression was paramount to the anarchist culture of discussion and neither does 'geography' appear as a controversial field that would warrant censoring. More importantly, even the private correspondence between anarchist geographers yields no sustained reflection on how to invigorate geography through anarchism. Both science and politics are being amply discussed, but few connections are made between the two. There are, moreover, hardly any traces of a conscious self-fashioning of this group of radicals and geographers as 'radical geographers' – whether with respect to fellow anarchists, nor in opposition to geographers of non-anarchist stripe. Reclus built a network of mutual support and collaboration and Kropotkin also worked tirelessly to procure scientific work for comrades, who like himself, were struggling to find regular employment in academia. But need and personal affinities (naturally often coinciding with political ones) might have been a more important motivation for the preference given to anarchists than a collective identity as belonging to the same epistemic community. If radical geography was indeed a thing, it was not one that required discussion or even definition. Reclus, Kropotkin, and their allies were not the ones to author manifestos in the manner of Harvey and Springer.

We need of course not stop at narrow definitions of either term, or of 'anarchist geography' for that matter. John P. Clark offers the following characterisation of Reclus that I find hard to disagree with: 'Reclus is the anarchist geographer par excellence. The term "anarchist geographer" captures perfectly the idea of his work: writing (*graphein*) the history of the struggle to free the earth (*Gaia*) from domination (*archein*)'.[35] Yves Lacoste rightly sees as one of Reclus' main contributions to geography his expansion of the scope of 'geographicity' (*géographicité*), that he made possible the introduction of notably the social and the political within the realm of geography.[36] This is the path continued by Springer who lauds geography's unboundedness and using a large and open understanding of his subject matter offers a spatial reading also of anarchist theorists who did not consider themselves geographers, for example Proudhon, Bakunin, and Goldman.[37] Other than perhaps the risk of losing disciplinary profile, I have no reservations against such a broadening of horizons. There is no reason to suspect that Kropotkin or Reclus would have. But the fact that they never came to discuss Proudhon, Bakunin, and Goldman as fellow geographers says much about their indebtedness to contemporary, rather than today's conceptions of geography and also about their ability to challenge such notions. Within these, there simply were no means for expressing an independent concept of an 'anarchist geography'. If such a category can be made a helpful analytical tool for scholars, this circumstance deserves to be pointed out.

The history of reception: radicals and the establishment

Another strand of intellectual history examines the impact of a thinker on other authors or a broader intellectual community, within their time and beyond. A strong programme of reception history understands texts as communicative acts and therefore regards the confrontation with the addressees as essential to arrive at possible meanings. To get a fuller appreciation of available interpretations of Reclus and Kropotkin, it is then promising to look not only at today's readings of anarchist geographies but also at the way their works were read by their contemporaries. Yet here we have a second problem – the anarchists' geographical writings, in their time, were not generally received as works of radicalism.[38] Not only were nineteenth-century geographer-anarchists reluctant to openly advocate 'anarchist geography', granted that there is a subversive anarchist colouring informing their work, it was so subtle it went almost unnoticed by contemporaries. The fact that they were highly appreciated by establishment geographers points to an underlying tension in the literature on anarchist geographers: on the one hand, their geography is presented as intrinsically revolutionary, on the other, their work is to have been highly influential. Unless we suppose that the bulk of nineteenth-century mainstream geography was crypto-radical, we cannot really have it both ways.

The latter position is far easier to back up with evidence. Although their ties were mostly informal, Reclus and Kropotkin had intense and good relations to the Geographical Societies of different countries.[39] Kropotkin had made his first explorations and publications under the aegis of the Russian Imperial Geographical Society; in the years before his involvement in the Commune, Reclus was on excellent terms with the Paris Société de Géographie. In the Geographical Societies of their native lands Kropotkin was offered the position of secretary[40] and Reclus was made member of different commissions.[41] These connections helped them when they were later imprisoned – renowned scientists petitioned for Reclus' release in 1871, and for Kropotkin's from the Peter and Paul Fortress in 1874, as well as from Clairvaux in 1883. Both in Russia and in France, Kropotkin obtained the permission to pursue his scientific work and have books sent to prison – not a likely scenario had such writings been considered explosive. Their increasingly open political involvement did not put an end to their cordial exchange with prestigious learned societies. Kropotkin, in fact, only started to build his connection to the Royal Geographical Society (the same society that would award its gold medal to Reclus in 1894) by the time he arrived in Britain as a known political exile and revolutionary.

Their intervention also demonstrates that the members of the Geographical Societies were under no illusions as to the political orientations of their anarchist associates. Their strategy of dealing with possible contradictions was twofold. Firstly, they simply refused to take their anarchism very seriously: John Scott Keltie, assistant secretary of the Royal Geographical Society, in 1913 regarded his friend Kropotkin as 'past [his] plotting days',[42] and Hugh Robert Mill, the society's librarian, commented sardonically on Kropotkin's *Memoirs of a Revolutionist*, 'you made a mistake on the title page – "Revolutionist" should have been "Geographer"'.[43] Secondly, they insisted on a clear separation between both activities: the same Keltie wrote in his obituary for Kropotkin that 'this is not the place to deal in detail with Kropotkin's political views, except to express regret that his absorption in these seriously diminished the services which otherwise he might have rendered to Geography'.[44] What is even more noteworthy is that the anarchists reciprocated these same two strategies: the tone Reclus adopts when addressing his sympathetic editor Paul Pelet is similarly jovial on politics, suggesting lightheartedly that Pelet might soon consider joining their camp.[45] Kropotkin reveals himself to Keltie as being able to abstract science from politics – when asked to give his opinion on the award of the RGS gold medal to the geographer Pëtr Semenov, he replied: 'Semenoff [*sic.*] is a Russian functionary and ready to serve under liberal and reactionary ruler alike and of course has no personal sympathy of mine, but scientifically, I think, [...] your choice was not bad.'[46] In his writings there is a noticeable distinction according to the audience targeted – writing on Darwin for *Le Révolté* he argued that 'the investigations of Darwin and his successors comprise [...] an excellent argument to the effect that animal societies are best organised in the communist-anarchist manner',[47] a formulation he deliberately left out when writing on the same subject for *The Nineteenth Century*. But such compromises seem not to have taken too high a toll for either side.

Sure enough, their connections were mainly to the more liberal members within these societies and publishing houses. They were however not exclusive to them. Halford Mackinder – hardly a suspect of anarchist sympathies – was acquainted with both Reclus and Kropotkin. He wrote a largely favourable review of Reclus' *Nouvelle Géographie Universelle*[48] and critically, but sympathetically discussed the proposition of a project for spherical maps by 'our distinguished brother geographer'.[49] Even more strikingly, he invited both Reclus and Kropotkin to contribute a volume to his projected *Regions of the World* series; Reclus on the Mediterranean, Kropotkin on the Russian Empire.[50] The volumes never materialised, but this was due to financial and temporal constraints rather than because of political disagreements as we might expect. Mackinder appears to have been unconcerned about his colleagues' political proclivities and this was also the case for the anarchists: the topic of the collaboration appears time and again in the correspondence between Reclus and Kropotkin, but there is never any mention of Mackinder's fervent defence of imperialism, British chauvinism or what would nowadays be called geopolitics. In their publications, Kropotkin and Reclus virulently attack the competitive vision of nature promoted by Huxley and Haeckel, but not Mackinder and 'geopolitics'. Indeed, Pelletier has pointed out that we find in Kropotkin and Reclus a description of central Asia that is not without similarities to Mackinder's famous 'heartland'-theory.[51] There is an epistemic resemblance between the geography of anarchists and 'classical' geopolitics that is acknowledged also by Lacoste who views Reclus as a proponent of 'benevolent geopolitics'.[52] Springer's suggestion of reading anarchist geography as 'anti-geopolitics' in the sense of Paul Routledge goes in a similar direction.[53] The anarchists, moreover, shared two other characteristic traits of geography as it took shape in the late nineteenth century: the idea that the human and the physical strand of geography could be brought together through an evolutionary narrative (that for imperialists and anarchists alike consisted of an amalgam of Darwinism with a Lamarckian twist)[54] and an emphasis on the propaedeutic suitability of geography – also in view of improving its standing at schools and universities.[55] From the point of view of epistemology then, Reclus and Kropotkin were hardly outsiders, but, to the contrary, in conversation with a field undergoing rapid transformation. That they appropriated these epistemic foundations for their own goals and filled them with a politically distinct content is not the same as saying that they founded an independent epistemology of their own that 'radically' challenged prevailing ones.

Intellectual biography: bringing anarchy and geography together

That neither Reclus and Kropotkin nor their contemporaries ever called their anarchism geographical or their geography anarchist imposes serious limits on the extent to which we can understand their combination as a radical project. At the same time, I have already expressed my discomfort with the view that treats the two as unrelated or even contradictory. This hesitation stems also from the extraordinary degree of confluence between both concerns in Reclus and Kropotkin's everyday life, practice, and identity. Biographers of Kropotkin or Reclus, even those who explicitly focus on either the geographer or the revolutionary, have never been able to completely shield off the other face of their protagonist;[56] I have already mentioned the role of collective biographies in substantiating the idea of an 'anarchist geography' in the nineteenth century. In view of its emphasis on praxis and the protean, non-systematic nature of anarchist thought, it has recently been suggested that biographical approaches are of particular relevance for the field of anarchist studies.[57] In the case of many anarchists, their outstanding life stories make for an intriguing subject in its own right. Life and thought can of course be brought together in the form of intellectual biographies that can give us a sense of the relative importance of certain ideas and their relationship to each other within the life of a thinker.

For the case of Reclus and Kropotkin, intellectual biography provides insights insofar as the chronological perspective that is typical for the genre brings to the fore turning points and conjectures in their engagement with anarchism and geography – it thus allows for a more dynamic view of their interlocking. All (auto-)biographical accounts retrospectively stress the predisposition of both passions from an early age on, even if only in the form of vague sentiment for the beauty of nature and a yearning for personal freedom and social justice. Although neither aspect appears in a very 'mature' form, Reclus' first essay carries a title that can be interpreted both geographically and politically – 'Développement de la liberté dans le monde' ('The Development of Freedom in the World', 1851).[58] For Kropotkin, it is the years spent in Siberia as a young official that he portrays as a crucial moment for his intellectual awakening as both a scientist and a socialist – seeing the misery and the inefficacy of the state made him 'prepared to become an anarchist'.[59] However, just like for Reclus, it took the experience of meeting the Bakuninists and the Jura Federation during the early 1870s to provide him with the vocabulary and practical experience of anarchism. What is more important for our purposes is that Kropotkin at this stage in his life believed that practicing geography diverted crucial energy from the revolutionary cause, that science was elitist like Bakunin had suspected; 'Science is an excellent thing. I knew its joys and valued them [...] But what right had I to these highest joys, when all around me was nothing but misery and struggle for a mouldy bit of bread.'[60] Even if the interest sprang up contemporaneously and from similar inputs, this is not in itself an argument for a necessary relation between geography and anarchy.

But their conception of either – or of their relationship – did not remain static. From a long-term perspective, we sense an increasing coming together of anarchy and geography in the course of their career. This initially translates into a continuous move away from a purely physical conception of geography: Kropotkin evolved from studies on orography and glaciation to a last uncompleted study of ethics; the Reclusian trilogy follows a succession from

La Terre (1868–1869), to *La Terre et les Hommes* (the subtitle of the Nouvelle Géographie Universelle (*NGU*), 1876–1894), to *L'Homme et la Terre*
(1905–1908). The last posthumously published series is the most obviously
political and at the same time most defying of geography's traditional area of
responsibility; its declared goal was to 'study man in the succession of the
ages as I had observed him [in the *NGU*] in the diverse lands of the globe and
to establish the sociological conclusions to which I have been led'.[61] For his
nephew and continuator Paul, it was only in this last oeuvre that Reclus tried
to bring anarchism and geography together – *L'Homme et la Terre* was supposed to 'affirm the unity of his perspectives of savant and anarchist, to develop
his book *Evolution et révolution* further at the same time as constituting a last
chapter to the *Nouvelle Géographie Universelle*'.[62] Reclus opened up to history
and sociology, whereas Kropotkin turned increasingly to biology: *Mutual Aid*
(1902), his cooperative reading of Darwin's theory of evolution, was conceived
of as a work of politics as much as science. Put simply, the more 'anarchist'
their science became, the less distinctively 'geographical' it was in name. But,
as conceded before, names only go so far. It remains to be seen what analogies
and differences there are between the widening of Reclus and Kropotkin's
interests and the broad scope for geography claimed by modern anarchist
geographers.

The history of science: nineteenth- and twenty-first-century geography

Modern interpretations of Kropotkin as radical geographer rarely point out
that, in later years, Kropotkin did in fact come to fully and openly associate
anarchy and science – and by that token, geography. In 1894 he stated that
'[t]he philosophy which is being elaborated by study of the sciences on the one
hand and anarchy on the other, are two branches of one and the same movement
of minds: two sisters walking hand in hand'.[63] In *Modern Science and Anarchism*
(1903) he defines anarchism as 'a world-concept based upon a mechanical
explanation of all phenomena, embracing the whole of Nature – that is,
including in it the life of human societies and their economic, political, and
moral problems'.[64] The notion of 'world-concept' of course has an evidently
geographical ring to it. Elsewhere, he describes the science of geography as 'a
philosophical review of knowledge acquired by the different branches of
science'[65]; recalling this broad definition of anarchism as 'a synthetic philosophy comprehending in one generalization all the phenomena of Nature'.[66]
Geography has, for Kropotkin, this same meta-scientific quality in that it
uniquely brings together natural and social science. While its social side is
crucial and corresponds with physical aspects, the primacy of the latter remains
incontestable. In his often-cited 'What Geography Ought to Be' (1885), Kropotkin
proposes a complete abolition of the classical humanities in schools. The
importance of geography resides in its capacity to assure the continuity of one
of the few aspects worth saving in classical education: 'Remaining a natural
science, [geography] would assume, together with history (history of art as

well as of political institutions), the immense task of caring about the humanitarian side of our education'.[67] Yet human geography is only one of four strands that he identifies in this essay. If, unlike *Modern Science and Anarchism*, 'What Geography Ought to Be' is quoted in every single modern appropriation of Kropotkin, it is not like passages that are chosen (his condemnation of colonialism and national self-conceit appear without fail). The abridged version published in *Antipode* in 1978[68] cuts out precisely those passages where he defines the physical branches of geography and argues for the replacement of Latin and Greek with 'exact sciences' alone. Clearly, the concerns of a modern and 'radical' geography only partially coincide with Kropotkin's.

While nineteenth-century and twenty-first-century anarchist geographers have in common a large definition of the scope of their subject, there are important distinctions to be made regarding their accentuation of certain aspects. These are not trivial; I believe they are indicative of larger epistemological differences in their understanding of geography and of its aims. The geographical vision of nineteenth-century anarchists is, in a sense, vastly more encompassing: Reclus especially is unique in his aspiration to bring together determinism with possibilism,[69] teleology with the idea of an open course of history, a care for minute detail and diversity with a general and systematising, truly world-geographical perspective. Twenty-first-century geographers interested in Reclus and Kropotkin, who without exception have their disciplinary home in human geography, have a tendency to stress those elements that resonate well with the concerns of their specialisation and time, where determinism, teleological narratives and the like have irrevocably come out of fashion. This is true also when it comes to judging the weight of individual arguments within the anarchists' world vision: disagreeable elements (Kropotkin and Reclus' failure to categorically condemn anarchist terrorism) or scientifically outdated residues (there are no appeals for reappropriating Kropotkin's Lamarckism) are given short shrift and are not treated as integral pieces of the puzzle. The more conveniently 'anarchist' aspects on the other hand are attributed a place of preeminence. Yet more problematically, even on the issues where Reclus and Kropotkin can be said to have prefigured anarchist geography in a contemporary sense, the literature is far from unanimous: Pelletier stresses Reclus' adherence to progressivism against widespread readings that align him with ecology also in its conservative dimension;[70] there is an unexpected amount of disagreement also concerning Reclus' position on (French) colonialism, especially for the case of Algeria.[71] Both sides in these debates find ample evidence to back up their case, making selectiveness a serious problem in an author who published as much as 30,000 pages – those who look for them, will have no difficulty finding in this monumental body of work passages echoing Eurocentric tropes and racial stereotypes that sound puzzling to modern ears. I do not mean to argue here that such instances are more revealing about Reclus' 'true' intentions, but the ambiguity even on such crucial points makes them a feeble epistemic foundation for 'anarchist geography'. The status of the anti-colonial dimension in Reclus' geography is hardly the same as that of post-colonial approaches – the (post-)colonial condition is not the *raison d'être*, nor the starting point of geography.

The epistemic relationship between science and politics that Kropotkin develops in *Modern Science and Anarchism* works inversely. He was more interested in transforming anarchism than he was in turning the epistemic foundations of geography upside down. It is not question here of founding an anarchist science – let alone in a Feyerabendian 'anything goes' sense – but of construing anarchism on the unshakable basis of inductive, scientific reasoning; '[anarchism's] method of investigation is that of the exact natural sciences'.[72] Their obsession with accuracy, as it emerges from Kropotkin's replies to letters to the editors[73] or Reclus' categorical repudiation of exaggerations in vistas, reliefs, or maps,[74] shows that the 'scientific' pretension in all their endeavours went beyond rhetoric. Kropotkin's notion of natural science is incontestably more comprehensive than our present one – 'human geography' would just as much be part of it. But the hierarchy is clear: when he transposes social science or anarchy to the realm of natural science, he proposes to adopt the latter's methods by reason of their supposed objectivity. This is a far cry from contemporary geographical theory, which Springer describes as being marked by an 'increasing tendency to liberate epistemological and ontological views from the illusion of disinterested objectivity'.[75] Unlike in Kropotkin's days, this scepticism about the neutrality of science is now a default position in the humanities. Historians have also had their share in this evolution – the history of science as a specialisation in its own right has for long been contributing to our understanding of science as a socially construed, context-dependent, and historically contingent enterprise. Such a view is as representative for our age as Kropotkin and Reclus' was for theirs: a historian of science today would naturally expect there to be fundamental differences between nineteenth- and twenty-first-century geography.

Context: science in the history of Russian political thought

In its relativising historicism and the interest in the evolution of epistemes, the history of science shares many of the methods and assumptions with the way the history of ideas has been carried out for the past half-century. The preoccupation with context, used in a variety of meanings,[76] has also become the trademark of intellectual history. For the history of science as well as for intellectual history, contextualism went hand in hand with a turn to politics – as a focus on the politics of science, or as the rise of the history of political thought, respectively. To approach the nexus between science and politics in anarchist geography also from this angle, I propose to contextualise and historicise the usages of science in political discourse. The positivist and materialist tradition in political thought is of course a long one. Reclus, who posits his understanding of science as 'the objective search of truth'[77] directly in opposition to religion, is very much its heir. Like Kropotkin, he was confident that science could provide a neutral, universal language uniting mankind on its path towards progress. But there is also a much more specific context that is highly illuminating for the particular type of association between radical politics and (geographical) science that we find in Reclus. For this, we have to bear in mind that virtually all of the members of Reclus' circle were Russians or had strong ties to Russia – even the Swiss Perron had spent several years in Russia and Élie Reclus worked as a correspondent for Russian journals. These ties have had graspable repercussions on their thought.

There are, first of all, noteworthy parallels between the geopolitical and geohistorical approach promoted by nineteenth-century anarchists and a tradition of geodeterminism in Russian historicism. Mark Bassin has devoted an important series of articles to the subject of how Russian thinkers – in a line stretching from the moderate conservative Sergei Solov'ev[78] (who is referenced in Reclus' *NGU!*[79]) to Marxist theoretician Georgii Plekhanov[80] – sought to account for Russia's backwardness by its environmental conditions. I cannot here delve into the substantial differences between the anarchist stance and others (the open plains of Asia are seen as a future potential rather than a historical hindrance), but there is another instance of discernibly 'Russian' roots to the thought of Reclus and Kropotkin: already in the late 1980s, Alexander Vucinich and Daniel Todes had offered materials for a Russian contextualisation of the interpretation of evolutionary theory put forth by the anarchist geographers.[81] Kropotkin's famous theory of mutual aid represents one of the most obvious intersections between science and anarchism in that it establishes solidarity as a law of nature thus producing a 'scientific' argument for morality without coercion by a state. The theory however becomes much less original when placed within the Russian intellectual context: reviewing the most influential Russian scientists of the period, Todes identifies a distinctive 'national style' that consisted of an unproblematic, sometimes enthusiastic reception of Darwin coupled with an uneasiness concerning the 'Malthusian bias' of the theory of evolution. From Slavophile conservatives to populist radicals, Russians sought to attenuate the bearing of the Malthusian metaphor of the 'struggle for existence' on the theory of evolution as a whole – Kropotkin's privileging of cooperative instincts over competitive ones was but one of the solutions proposed. For Vuchinic, who has made very similar observations on the reception of Darwin in Russia, this similarity to other Russian approaches indicates that 'in the last analysis, Kropotkin was a *Russian* anarchist'.[82]

If we accept that such readings of 'Darwin without Malthus' or historical geodeterminism are a Russian particularity, their prevalence across the political spectrum means we have to look further when making a case for them as an anarchist particularity. We can do so by linking them to the broader intellectual context of nineteenth-century Russia, which is characterised by an exceptionally close intertwinement of radicalism, often of an anti-statist nature, with science. In the era of political repression of philosophy and liberal subjects at universities, many of Russia's political thinkers were scientists by training. Aleksandr Herzen, for example, a lifelong friend and correspondent of Bakunin and with his avocation of communitarian socialism and individual liberty a proto-anarchist in many ways, began his publishing career with essays like 'Dilettantism in Science' (1842) and his 'Letters on the Study of Nature' (1845–1846). The cloak of science made it possible to express criticism in face of censorship, but it also meant that distinctions between science and politics were increasingly blurred. The generation of the intelligentsia following Herzen, to which the Russian anarchist geographers age-wise belonged, are often dubbed 'nihilists'.[83] Nihilists took the idea of science put to the use of social progress to its extremes, further even than comparable forms of materialism in Western Europe. James P. Scanlan has described Russian materialism as 'less a precisely articulated ontological position than a grand, science-worshipping worldview that sought to undermine both religion and the state'.[84] In the nihilist movement, radicalism and scientism crystallised in a unique manner. Kropotkin describes how his brother abandoned his passion for poetry under the spell of nihilism in order to 'plunge headlong into the natural sciences'.[85] Kropotkin's formulation of anarchism as science certainly drew on this intellectual climate; his holistic vision of science is comparable to that of leading nihilists like Nikolai Chernyshevskii.[86] At different occasions, Kropotkin himself acknowledged the proximity of nihilism to anarchism.[87]

As this kind of scientifico-political discourse travelled with the Russian anarchists into Western European exile, it was deployed in the context of a different debate. Kropotkin uses the claim to the scientific nature of anarchism as an attack on the anarchists' main opponents: he accuses Marx of idealist 'metaphysicism', his political economy is but a pseudo-science.[88] Anarchy's strength derives from its being anchored in exact natural science rather than abstract speculation. Why, then, when following their forefathers into their crusade against Marxism, do present-day anarchists desist from adapting this readily available argument?

Intersecting genealogies: Marxist and anarchist histories of geography

Springer understands the historical connection between anarchism and geography as an on/off relationship – 'there have been periods of deep engagement and connection, and times when ambivalence and even separation have occurred'.[89] The genealogy of anarchist geography is, I believe, more complex. It is necessary to distinguish in the thought of early anarchist geographers between different 'geographical' strands, which were each to reach a largely different intellectual destination. The argument of anarchism as science proved to have very little posterity. Not only was it hardly accepted by scientists at the time, even in anarchist circles it immediately met with criticism: Errico Malatesta lamented that Kropotkin 'put all his social aspirations on science', and because of his confidence in the scientific method, 'fell into a mechanical determinism that seems even more paralysing' for the anarchist cause.[90] In the decades following the First World War, anarchist thought evolved more in Malatesta's sense. By the 1960s at the latest, it came to self-consciously embrace its utopian, visionary dimension at the same time as the positivistic quest for laws governing history and society (which both Reclus and Kropotkin believed was feasible) increasingly lost credibility, throughout the social sciences and on the left especially.

The elements that Springer regards as the essence of anarchist geography, its focus on practice in particular, have the same resonance with the schemes of the New Left and its contemporary renaissance in the Occupy movement. Springer and like-minded geographers see 'direct action, civil disobedience, and Black Bloc tactics; in the communes and intentional communities of the co-operative movement; amid DIY activists and a range of small-scale mutual aid groups, networks, and initiatives'[91] as constitutive for anarchism in geography, or rather, as spatial practices in anarchism. It is indeed possible to trace, as Springer and Anthony Ince do, such an emphasis on decentralisation and bottom-up organisation back to Kropotkin.[92] No matter that Kropotkin never labelled these writings as 'geography', the argument for local action is present in *Mutual Aid*, in the *Conquest of Bread* (1892) as well as *Fields, Factories and Workshops* (1898). The reason I do not develop on this type of evidently geographical reasoning here is not that I believe it to be irrelevant, but because the task has already been admirably done by contemporary anarchists. The lasting legacy of this theorising is confirmed by the fact that there is a direct and documentable genealogy. There is an uninterrupted continuity in the reference to Kropotkin all through the twentieth century, stretching from Patrick Geddes to Colin Ward and Murray Bookchin up to contemporary social theorists like James C. Scott or David Graeber. As city-planners, anthropologists, and community developers began to explore the anarchist heritage, its spatial dimension was continuously reinforced and made more explicit.

Our understanding of space, however, fundamentally evolved in the last quarter of the twentieth century. In France and in America influential authors like Henri Lefebvre, Edward Soja, Michel Foucault, Doreen Massey and not least David Harvey came to regard space as a socially 'produced' entity;[93] it was no longer thought of as a mere container for human action, but as the result of social interactions, as a construed formation. Space freed from its purely material(ist) definition became more dynamic and abstract. Yet the problem for our case of the anarchist spatial tradition is that the proponents of the radical turn in geography (and possibly, the spatial turn in radicalism?) overwhelmingly identified with the Marxist tradition. This tradition, as they all acknowledged, had thus far shamefully neglected space – Lacoste speaks of 'the silence, the "blank" with regard to spatial problems' in the work of Marx.[94] In its exclusive focus on time, Marxism had failed to take notice of distinctively 'spatial' issues like difference and diversity that urbanists, sociologists, and geographers now sought to introduce into the framework of Marxist theory. Space, in short, became a means of liberating Marxism from its increasingly burdensome Eurocentric universalist legacy. For this spatialisation of Marx, theorists however hardly looked to the anarchist geographers; Lacoste, who went perhaps furthest in turning his back on Marxism, is highly exceptional in this sense. There is no genealogy to be established here: Lefebvre cites Reclus' depiction of the Pyrenees,[95] but does not treat him as a major intellectual influence. Kristin Ross credits Reclus as one of the inventors of 'social space', but this feels more like a superposition of Lefebvre on Reclus than *vice versa*.[96]

Transforming Marxist theory through the expansion of a spatial dimension, theorists of the spatial turn did pick up much of the anarchist critique, even if more or less unknowingly. Harvey is one of the few to acknowledge this similarity;[97] still, his engagement with Reclus and Kropotkin remains quite superficial. They are, for him, proponents only of the decentralising, particularising strand that Springer and Ince underline. In so doing, both Marxists and anarchists underestimate the presence of a universalising strand in Reclus and Kropotkin, that is, the integration of their localism within a larger, holistic perspective. As is the case for most of the movements of the ilk of Occupy, for Kropotkin and Reclus, the local dimension could not stand on its own. Both were critical of isolationism in Fourierist *phalanstères* or other autonomous communities.[98] For Kropotkin, the demise of the medieval city guild could be explained by its failure to integrate the surrounding countryside within their egalitarian but strictly internal social organisation.[99] But he saw no reason why a future society should not be able to expand the principles of self-organisation and mutual aid to a larger, possibly even global scale. This internal necessity of expansion along federative lines opened the door to globalising narratives – a perspective that, given the universalism of their scientific anarchism, Reclus and Kropotkin had no reservations against. Springer recognises this coexisting generalising endeavour, but finds it not worthy of much discussion, commenting only that 'the universalism of [Reclus'] thought has become unfashionable as a result of poststructuralism's influence on the academy'.[100] When Harvey elsewhere identifies a conflict between universalising narratives and their tendency towards 'flattening out all geographical differences',[101] he does not note that Reclus and Kropotkin were struggling with the exact same tension: they never quite achieved the reconciliation between their sensitivity to diversity, to local, individual initiative and the kind of synthesising teleology that they aimed to construe based on their faith in science and their holistic approach to geography. If the anarchist and the Marxist strand in geography have come to interact, each side remains reluctant to openly endorse the other. Marxist geographers have failed to recognise the affinity of their ambitious project of combining the universal and the particular with that of nineteenth-century anarchists. Modern anarchists, on the other hand, hardly assume the indebtedness of their modern conception of space – the very notion that allows them to put forth 'geographical' readings of a wide range of thinkers and movements – to the (Marxist) thinkers of the spatial turn.

Conclusion

In both the nineteenth century and the present, anarchist geographers see their activities as naturally compatible, if not complementary. With their discipline-related depreciation of space, historians have long failed to fairly assess the bearing of the association. However, if we accept that for Reclus, Kropotkin, and their colleagues and comrades, anarchy and geography were meaningfully related, we have to take seriously their understanding of geography in all its historical dimensions. The broad scope of their 'geographical' reach allows for many interpretations: radical geographers have understandably been most interested in how Reclus and Kropotkin's work resonates with present concerns, such as debates on globalisation, ecology, community activism, the anthropocene, etc. For Ferretti, Reclus anticipated Dipesh Chakrabarty, Jack Goody, Jared Diamond, and many more contemporary historian-theorists at once.[102] Such associations also testify that there is an interest in reappropriating Reclus as a (world-)systematic, large-scale thinker. However, present-day geographers stop short of reclaiming such aspects with their full implications. This is, firstly, because it comes with the kind of holistic and materialist worldview that has a terribly outdated feel to it. The radicalism of such a view, of 'science' as a neutral force of progress and enlightenment, is difficult to appreciate from the point of view of a twenty-first century epistemology – yet anarchist geographers never quite leave the premises of their age. It is, secondly, for political reasons as the 'big picture' view is at risk of bringing anarchist geography in dangerous proximity to Marxist approaches.

Such readings of historical geography through the lens of present concerns are not always innocuous. Whereas Springer's geography is openly radical, he consults 'history' not in a consciously militant way but as a more or less neutral, factual background to support his geographical argument. History thus becomes a smoke screen for a political agenda: for all his references to history, the issue of 'anarchist geography' remains for him a *political* problem – a case of 'anarchist' against Marxist, conservative, imperialist, or quantitative geography. The opposition to Marxism in particular is, in my view, a false dichotomy. It poses a pseudo problem, for the matter is essentially *historical* in nature: in the nineteenth century, anarchists combined a universal synopsis, embedded in the evolutionary perspective that was typical for their time, with a sensitivity for the variety of human life across the earth. Incidentally, I believe it is in this first totalising dimension where Kropotkin and Reclus make the strongest case for a connection between geography and anarchy. The decline of grand narratives in the twentieth century had severe repercussions on Marxist and anarchist thought alike: Marxist theorists sought to incorporate the diversifying dimension that their theoretical frame had hitherto lacked; crucially, this often took place with the explicit reference to space – a connection that nineteenth-century anarchists had a premonition of, even if perhaps less consciously. Anarchist geographers today tend to focus exclusively on this last centrifugal aspect, deploying it also as a critique of Marxism's centripetal universalism. In downplaying the affiliation of both strands, they – quite ironically – underestimate the extent to which nineteenth-century anarchists can indeed be seen as congenial to the wider tradition of radical geography. If instead, the goal is to seclude an 'anarchist geography' from the field of radical geography, this would imply breaking with a large part of the historical heritage of anarchist geographical thought.

From the point of view of a historian, the usages of history in radical geography remind one of Malatesta's critique of Kropotkin's use of science:

> Nonetheless, it seems to me that he lacked something to be a real man of science: the ability to forget his desires, his prejudices to observe the facts with dispassionate objectivity. [...] Whatever the conclusions he could draw from contemporary science [...] he would have remained anarchist in spite of science.[103]

Historians have a reputation as killjoys and as being too little concerned with the wider present-day relevance of their investigations, happy to consign their subjects to the famous ash heap of history. The many routes opened up for anarchist geography by recent contributors to the field are doubtlessly more inspiring and I gladly leave the terrain of renewing anarchist geography to anarchist geographers. But as a historian interested in the thought of Kropotkin and Reclus, I feel they deserve more than hagiography. Given the ambitious programme of radical geography as well as its heartfelt desire to engage with history, I find its reluctance to deal with the obvious divergences between geographies separated by a century a bit disappointing. From an anarchist point of view, far-reaching (epistemological) differences to nineteenth-century founding fathers ought not to be a problem since anarchism self-consciously assumes its protean and dynamic nature. If Reclus and Kropotkin are not to be decorative props, but an actual source of inspiration, a critical engagement with the more controversial aspects in their thought could turn out to be just as productive for a thorough recasting of geography. It then becomes possible to remain an anarchist not in spite of, but with history.

Notes

1 I would like to thank Federico Ferretti, David Motadel, Jürgen Osterhammel, Lucian Robinson, and Nadine Willems for comments on earlier drafts of this chapter, as well as the participants of the 'Historical Geographies of Anarchism' panels at the Royal Geographical Society Annual International Conference in Exeter in September 2015. The opinions expressed in this chapter, as well as all remaining mistakes obviously, are entirely my own.
2 Springer, 'Anarchism and Geography. A Brief Genealogy of Anarchist Geographies'; Springer, 'Why a Radical Geography Must Be Anarchist'; Springer, 'The Limits to Marx. David Harvey and the Condition of Postfraternity'; Springer, *The Anarchist Roots of Geography. Toward Spatial Emancipation*.
3 Harvey, 'On the History and Present Condition of Geography', 8.
4 Ferretti, 'The Correspondence Between Élisée Reclus and Pëtr Kropotkin as a Source for the History of Geography'; Ward 'Alchemy in Clarens. Kropotkin and Reclus, 1877–1881', 209–226.
5 Springer and Harvey concur that anarchism also played an important and nowadays underappreciated role in the early days of the formation of radical

geography in the 1960s; Springer, 'Anarchism and Geography', 51–52; Harvey, '"Listen, Anarchist!" A Personal Response to Simon Springer's "Why a Radical Geography Must Be Anarchist"'.

6   Lacoste, 'Hérodote et Reclus'.
7   Springer, 'Anarchism and Geography', 51.
8   Springer, 'Why a Radical Geography', 250.
9   Pelletier, *Géographie et anarchie. Reclus, Kropotkine, Metchnikoff* ; Ferretti, *Élisée Reclus. Pour une géographie nouvelle*; Schmidt di Friedberg, *Élisée Reclus. Natura ed educazione*; Lefort and Pelletier (eds), *Élisée Reclus et nos géographies*.
10 Cf. esp. the special issue 'Élisée Reclus' of *Hérodote* 22(1981).
11 One of Pelletier's chief concerns is to keep Reclus at bay from 'ecologism', Pelletier, *L'imposture écologiste*, 51–55. Ferretti focuses on anti-imperialism and post-colonialism: Ferretti, 'They Have the Right to Throw Us Out'.
12 For a direct reference, Ferretti, 'De l'empathie en géographie et d'un réseau de géographes. La Chine vue par Léon Metchnikoff, Élisée Reclus et François Turrettini'.
13 Pelletier, *Géographie et anarchie*, 78.
14 Ferretti, *Élisée Reclus*, 408.
15 According to Harvey, Reclus' work 'is seriously flawed by the absence of any powerful theory of the dynamics of capitalism,' Harvey, 'On the History', 8–9.
16 Harvey, '"Listen, Anarchist!" A Personal Response to Simon Springer's "Why a Radical Geography must be Anarchist"'.
17 Harvey, 'Listen, Anarchist!'
18 Reclus and Kropotkin have by now found their ways into standard histories of geography, though often in a marginal position. Livingstone, *The Geographical Tradition*; Berdoulay, *La formation de l'école française de géographie*.
19 Dunbar, *Elisée Reclus. Historian of Nature*; Jud, *Elisée Reclus und Charles Perron, Schöpfer der 'Nouvelle Geographie Universelle'*; Fleming, *The Geography of Freedom. The Odyssey of Élisée Reclus*; Clark and Martin, *Anarchy, Geography, Modernity: Selected Writings of Elisée Reclus*.
20 Kropotkin appears prominently in Eltzbacher, *Der Anarchismus*; Crowder, *Classical Anarchism*; Woodcock, *Anarchism. A History of Libertarian Ideas and Movements*; Joll, *The Anarchists*.
21 Woodcock and Avakumović, *The Anarchist Prince: A Biographical Study of Peter Kropotkin*; Pirumova, *Pëtr Alekseevich Kropotkin*; Miller, *Kropotkin*; Cahm, *Kropotkin and the Rise of Revolutionary Anarchism, 1872–1886*; Kinna, *Kropotkin. Reviewing the Classical Anarchist Tradition*.
22 Sokolov, 'Pëtr Alekseevich Kropotkin kak geograf'.
23 This – in spite of its title – is the case even of Morris, *The Anarchist Geographer. An Introduction to the Life of Peter Kropotkin*. On 'science' more broadly Dugatkin, *The Prince of Evolution. Peter Kropotkin's Adventures in Science and Politics*; Confino and Rubinstein, 'Kropotkine savant'; Adams, *Kropotkin, Read, and the Intellectual History of British Anarchism*.
24 MacLaughlin, *Kropotkin and the Anarchist Intellectual Tradition*.
25 Miller, *Kropotkin*, 134; Woodcock makes a similar argument, Woodcock, *The Anarchist Prince*, 227.
26 Koselleck, 'Einleitung', in Koselleck, Conze and Brunner (eds), *Geschichtliche Grundbegriffe. Historisches Lexikon zur politisch-sozialen Sprache in Deutschland*.
27 Reclus, *L'Homme et la Terre*, vol. 2, 112; vol. 3, 261; vol. 3, 428; vol. 4, 147; vol. 4, 515; vol. 4, 225–231.
28 Prudovsky, 'Can We Ascribe to Past Thinkers Concepts They Had No Linguistic Means to Express?'
29 Blunt and Wills, *Dissident Geographies: An Introduction to Radical Ideas and Practice*, 2.

30 Letter Onésime to Élisée Reclus, 11 April 1871, cited in Pelletier, *Géographie et anarchie*, 131.
31 Ferretti, *Élisée Reclus*, 116, 128–130.
32 Letter to Élisée to Fanny Reclus 8 July 1871, Bibliothèque Nationale de France, NAF 22913, 62.
33 Alavoine-Muller, 'Un globe terrestre pour l'Exposition Universelle de 1900. L'utopie géographique d'Élisée Reclus'.
34 This claim, picked up by Woodcock and many others in his wake, is only traceable to a few lines, written over three decades after the events in the memoirs of Kropotkin's friend John Mavor (Mavor, *My Windows on the Street of the World*, 75). It seems more likely that Kropotkin sought employment in the University Extension Programme.
35 Clark, *Anarchy, Geography, Modernity: Selected Writings of Elisée Reclus*, 61.
36 Lacoste, 'Elisée Reclus, une très large conception de la géographicité'.
37 Springer, 'Anarchism and Geography', 49–50.
38 For minor nuances: Ferretti, *L'Occidente di Élisée Reclus. L'invenzione dell'Europa nella Nouvelle Géographie Universelle, 1876–1894*, 164, 328.
39 Kearns, 'The Political Pivot of Geography'.
40 Kropotkin, *Memoirs of a Revolutionist*, vol. 2, 16.
41 Jud, *Élisée Reclus und Charles Perron*, 26–28.
42 Letter J.S. Keltie to P. Kropotkin, 15 December 1913, Archives of the Royal Geographical Society, CB 7.
43 Letter H.R. Mill to P. Kropotkin; Gosudarstvennii Arkhiv Rossiiskoi Federatsii [hereafter GARF], f. 1129, op. 3, d. 287, l. 37.
44 Scott Keltie, 'Obituary'.
45 Letter E. Reclus to P. Pelet, 2 January 1881, Bibliothèque Nationale de France, Nouvelles Acquisitions Françaises, 16798, 7.
46 Letter P. Kropotkin to J.S. Keltie, 1 May 1897, RGS, CB 7.
47 Cited in Todes, *Darwin Without Malthus. The Struggle for Existence in Russian Evolutionary Thought*, 130–131.
48 Mackinder, 'Reclus' Universal Geography'.
49 Mackinder et al., 'On Spherical Maps and Reliefs: Discussion', 294.
50 Mackinder, *Britain and the British Seas*, frontmatter.
51 Pelletier, *Géographie et anarchie*, 287–288; Ferretti, *Élisée Reclus*, 295, 336.
52 Lacoste, 'Elisée Reclus, une très large conception'; Kearns, *Geopolitics and Empire. The Legacy of Halford Mackinder*, 86–90.
53 Springer, 'Anarchism and Geography', 53.
54 Livingstone, *The Geographical Tradition*, 187–189; on Kropotkin's Lamarckism, 256–257.
55 Stoddart, 'The RGS and the Foundations of Geography at Cambridge'; Ferretti, 'Géographie, éducation libertaire et établissement de l'école publique entre le 19e et le 20e siècle'.
56 Early biographical accounts authored by anarchist acquaintances and family members are the only ones to give equal weight to both strands, Ishill, *Elisée and Elie Reclus: in Memoriam*; Nettlau, *Elisée Reclus. Anarchist und Gelehrter (1830–1905)*; Reclus, *Les frères Élie et Élisée Reclus. Ou du protestantisme à l'anarchisme*.
57 Berry and Bantman (eds), *New Perspectives on Anarchism, Labour and Syndicalism. The Individual, the National, and the Transnational*, 5–9.
58 Reclus, 'Développement de la liberté dans le monde', 1851, first published in *Le libertaire* in 1925, available at Internationaal Instituut voor Sociale Geschiedenis, ZF10170.
59 Kropotkin, *Memoirs of a Revolutionist*, vol. 1, 251.
60 Kropotkin, *Memoirs*, vol. 2, 20.

61   Reclus, *L'Homme et la Terre*, vol. 1, i.
62   Reclus in the afterword to Reclus, *L'Homme et la Terre*, vol. 6, 567.
63   *Les Temps Nouveaux*, 1894, cited in Cahm, *Kropotkin*, 2.
64   Kropotkin, *Modern Science and Anarchism*, 53.
65   Kropotkin, 'On the Teaching of Physiography', 359.
66   Kropotkin, *Modern Science and Anarchism*, 53.
67   Kropotkin, 'What Geography Ought to Be', *The Nineteenth Century, A Monthly Review*, 946.
68   Kropotkin, 'What Geography Ought to Be', *Antipode* 10(1978): 6–15.
69   Arrault, 'La référence Reclus. Pour une relecture des rapports entre Reclus et l'Ecole française de géographie', 1–14.
70   Pelletier, *Géographie et anarchie*, 323–367.
71   The literature on the topic is vast: Soubeyran (*Imaginaire, science et discipline*) Baudouin ('Reclus a colonialist?') and Giblin ('Reclus et les colonisations') see Reclus as a proponent of colonisation in a civilising, humanitarian sense. Pelletier (*Albert Camus, Élisée Reclus et l'Algérie. Les 'indigènes de l'univers'*) and Ferretti ('They Have the Right to Throw Us Out') reject this position. The most detailed and nuanced study for the Algerian case is Deprest, *Élisée Reclus et l'Algérie colonisée*.
72   Kropotkin, *Modern Science and Anarchism*, 53.
73   GARF, f. 1129, op. 2, d. 74.
74   Reclus, 'L'enseignement de la géographie'; Reclus, 'On Spherical Maps'.
75   Springer, 'Anarchism and Geography', 54.
76   Burke, 'Context in Context'.
77   Reclus, *L'Homme et la Terre*, vol. 6, 387.
78   Bassin, 'Turner, Solov'ev, and the "Frontier Hypothesis". The Nationalist Signification of Open Spaces'.
79   Reclus, *Nouvelle Géographie Universelle. La terre et les hommes*, vol. 5, 896.
80   Bassin, 'Geographical Determinism in Fin-de-siècle Marxism: Georgii Plekhanov and the Environmental Basis of Russian History'.
81   Todes, *Darwin without Malthus*; Vucinich, *Darwin in Russian Thought*.
82   Vucinich, *Darwin*, 346.
83   Lovell, 'Nihilism, Russian', in Craig (ed.), *Routledge Encyclopedia of Philosophy*; Cassedy, 'Nihilism in Russia and as a Russian Export', in Horowitz (ed.), *New Dictionary of the History of Ideas*; Schmidt, *Nihilismus und Nihilisten. Untersuchungen zur Typisierung im russischen Roman der zweiten Hälfte des neunzehnten Jahrhunderts*.
84   Scanlan, 'Russian Materialism. The 1860s', in Craig (ed.), *Routledge Encyclopedia of Philosophy*.
85   Kropotkin, *Memoirs*, vol. 2, 111.
86   Woehrlin, *Chernyshevskii. The Man and the Journalist*; Paperno, *Chernyshevsky and the Age of Realism. A Study in the Semiotics of Behavior*.
87   Kropotkin's entry on 'Nihilism' for the *Chamber's Encyclopaedia*, GARF, f. 1129, op. 1, d. 459, l. 8.
88   Kropotkin, *Memoirs*, vol. 2, 192.
89   Springer, 'Anarchism and Geography', 46.
90   Malatesta, 'Sur Pierre Kropotkine. Souvenirs et critiques d'un de ses vieux amis'.
91   Springer, 'Anarchism and Geography', 47–48.
92   Springer et al., 'Reanimating Anarchist Geographies. A New Burst of Colour'; Ince, 'In the Shell of the Old: Anarchist Geographies of Territorialisation'.
93   The foundational reference is Lefebvre, *La production de l'espace*.
94   Lacoste, *La géographie, ça sert d'abord à faire la guerre*, 96.
95   Lefebvre, *Pyrénées*.
96   Ross, *The Emergence of Social Space. Rimbaud and the Paris Commune*, 91.

97   Harvey, 'Listen Anarchist!'
98   Pelletier, *Géographie et anarchie*, 508–514.
99   Kropotkin, *Mutual Aid. A Factor of Evolution*, 189–190.
100 Springer, 'Anarchism and Geography', 49.
101 Harvey, 'Cosmopolitanism and the Banality of Geographical Evils', 535.
102 Ferretti, 'De l'empathie'.
103 Malatesta, 'Sur Pierre Kropotkine'.

PART III

Anarchist geographies, places and present challenges

8 Lived spaces of anarchy
Colin Ward's social anarchy in action

David Crouch

Introduction

The late Colin Ward was a major figure in British anarchist thinking and action, and was also influential in Italian anarchist circles. He believed that anarchy emerges and dwells in human life and relations and, though he would never use such a rampage of an expression, human/other-than-human relations. These he often investigated and articulated through human-space relations and their creativities; creativity was clearly at the heart of what for him anarchy involved, that he found in numerous everyday and so-called mundane life. His geographies of anarchy were the spaces of everyday life. He co-edited the British anarchist journal *Freedom* in the years 1947–1960, then to be known as *Anarchy*, which he co-founded, between 1961 and 1970. In the latter publication he deployed his liking for strong and striking artwork in the use of contemporary graphical illustrations on their numerous dynamic front covers.

In this chapter I seek to connect Ward's ideas and their origins with the very pragmatic, practical approach to spaces of people's lives. Certainly, his spaces were not the barricades but discovered in a multiplicity of very diverse everyday creativity. For two years, Colin and I researched and co-authored, and in that practice I felt I was able to engage closely his ideas and approaches to human possibility, i.e. the potentiality of every human being to contribute to the value and meaning of their lives.

Life, living and lived spaces provided the sustaining focus of his work,[1] in which he found everyday anarchy in action, in the potentialities and changing identities through creativity in non-heroic ways of doing, thinking, being, relating and much more. In this constructive insight into lived anarchy, Ward turned away from anarchisms' familiar emphasis on resistance, where anarchy was presumed to require *a priori* the clearing away of blockages to opportunity by institutions and corporations. For example, housing provided a familiar scope for action, most notably in Colin's enthusiasm for self-build as an alternative to market or state control. He joined up with a city planner, Denis Hardy in their project on *Arcadia for All*.[2] This work considered self-made homes to be not makeshift but self-built and self-enhanced – homes, familiarly around the British coasts in the form of 'plotlands', a great opportunity in the depressed 1930s and after the Second World War for low/no income house-holds to build their own.

More widely, his anarchistic spaces were to be found in the city, the urban and the village. The geographer Chris Philo pointed to Ward's insights on the child in the country as shining a light on the often neglected constraints but also potentiality and constraints for spaces and their human action.[3] These contributions had a strong recognition of and respect for 'play', again in the vein of Ward's thinking on freedom to do, to influence, to affect. Play was something he valued and regarded as may be done by adults as much as children: something we might today label as leisure, but also creativity and freedom.[4] Moreover, he left important insights that I am sure he would have presented to contemporary academics and their masters in the insidious institutional commercialisation of our universities. His central philosophy was that there exist, already and across the world, numerous examples of anarchy in action. Yet Ward also realised the importance of changing systems, the ways in which society was administered, as in his case for a more anarchy-informed way that could have been possible, given the will, in the way in which the UK's National Health Service came to be established in the middle of the last century.[5]

Colin Ward's enthusiasm was for freedom rather than resistance; a positive shift in much conventional anarchy thinking, and engaged ideas of practical, and already existing action that revealed numerous further potentialities, and the sources of their potential. In his book *Influences* he documented, reflec-tively in non-dogmatic and unpretentious language, philosophies of creative and constructive dissent that had helped him construct what came to be a unique path in contemporary anarchist thought-action for sixty years until his death in 2010.[6] The unexpectedly diverse thinkers include W. Godwin and M. Wollstonecraft on education; A. Herzen on politics, the geographer P. Kropotkin on economics, considered in a way that breaches its disciplinary boundary and connects society, human relations and culture. He follows these with M. Buber on society. Two further pairs of influence concerned W. Lethaby and W. Segal, very different architects but sharing an interest in individuals' potential directly to affect their immediate surroundings. Perhaps most surprising of all, he completes his main influences with two individuals in planning, a sphere not known for or familiar with anarchist ideas. They are the British pioneer of garden cities Patrick Geddes and the American anarchist thinker Paul Goodman. Typically always eager to connect thought with one's own action, almost as a priority, at the end of these reflections he takes the reader towards not simply 'Further Reading' but 'Read for Yourself'.[7]

In this chapter, next I unpick the ways in which Ward took and worked on these influences, most notably those of Kropotkin, as these seem more central than others. Such discussion is followed by a closer inspection on where Colin's thought always led to action.[8] That thought persistently became informed through understanding the action. This chapter is, with relief, not of my autobiographical engagement with Colin, yet in part I touch upon how I learnt from two enriching years and more of researching and writing together; interviewing and giving talks. In passing I engage something of Ward's reiterative mutual attention to philosophy and practice. He was the most inspiring colleague in my academic career, along the way bordering onto my politics too. Whilst he was gentle in his thinking and conversation around human action he was also strong; humble in talking, unequivocal and consistent in his inspiration.

Influences and ideas

Ward's attachment to the idea of mutual aid was directly influenced by his enthusiasm for Kropotkin's elucidation of the idea, in his book of the same name. Whilst Kropotkin was aligned also with communist overthrow of the state and its institutions, Ward positioned this in the light of the times in which he lived: the easy submission of human beings to starvation in a time of reliance on imported food, rather than their self-production; when the ever increasing scale of food production and the dehumanisation of labour led to extreme hardship. Redistributing factories amongst fields of co-operatively produced food provided, for him, a key manipulation of space into local control. The internationalism of his thinking acknowledged the power, into the twentieth century, of control over how things were done at home and extended to how the manner of farming in Britain destabilised especially colonial countries in their requirement to produce food for export. Discussing these views, and in the light of the emergent realities following the Russian revolution, Ward emphasised the way that Kropotkin felt at the beginning of the Soviet period: the inevitable failure of asserting a central authority on far reaches across an empire and directing attention to federations of independent units.[9]

Ward's considerations of 'how to get there' were tempered through his engagement with Martin Buber, who he acknowledged to be 'a model of gentle benevolence'.[10] This became clear in his reading of Buber's *I and Thou* book,[11] on forms of human engagement. For Buber, Kropotkin's approach was welcome but for its means, risking the threatening chasm of the unknown. The answer would be not an imagined, hoped-for instant utopia, but through working on other forms of human and social relationships. Ward worked for forty years identifying and unpacking what he found to be relevant forms of human and social relationship in our contemporary times. He acknowledged their imperfections, their often convoluted evolution. For him, these were reality, and remain so. These for him demonstrate potential in what happens now.

As his philosophy from Kropotkin reasoned, these relations featured great strength from the distinctive combination of mutual aid and self-reliance. Buber emphasised these too, as central to the sought for human condition. Ward also felt that the conflict between the social and the political was a permanent aspect of that condition. Perhaps, there is a crude duality between the social and the political which betrays a misunderstanding. Values, attitudes and the meanings of things significantly emerge not merely in the form of pre-givens but through our lives, our doings, actions and close-up human relations, their occurrence through our actions and human relations I have called 'gentle politics'.[12] Their character merges the social and the political.

Ward's response was that

> to weaken the state we must strengthen society, since the power of one is the measure of the weakness of the other. Buber's exploration of the paths to utopia, far from confirming an acceptance of the way things are, confirms, as do several of (his) influences, that the fact that there is no route map to utopia does not mean that there are no routes to more accessible destinations.[13]

Ward pursued and honed these influences in the way he engaged individuals and families building their own homes, something particularly possible where vacant sites are small and unprofitable for developers. He was a pioneer supporter of self-build, DIY housing, avoiding the costs that leak away to middle men.[14] He worked hard in his writing, giving talks and in conversation to encourage schools to engage children more intently in cultivating food and participating more enquiringly in their surroundings, close-up environments, as means to connect with life around them.

In ways, these examples, exemplars and expressions of mutual aid amidst self-reliance participate in adjusting social and human relationships. They render a market where you can give and take; where the currency is co-operation.

Ward has been criticised for the way in which his focus on 'the local' of small worlds can be construed as inward, parochial and a matter of closure.[15] Yet he strenuously argued that the local, in a sense, does exist everywhere and can simultaneously with overlapping networks of a federal arrangement; always fluid, open and balanced between probably conflicting undercurrents calmed through careful mutual attention. Accusations of being 'too local' are ill placed. We live in a shared world, in whatever other condition we live. It is too easy to feel the local to be small minded or isolationist. To argue so is to fail to acknowledge the achievability of making adjustments to social and political relations at all. To reach too widely is to reach domination. A flexible, fluid and open approach achieves those numerous adjustments that Ward, taking Buber a step forward, identified as the numerous modest steps to where we might reach.

In *Anarchy in Action* he writes:

> The argument of this book is that an anarchist society, a society which organises itself without authority, is always in existence, like a seed beneath the snow.... far from being a speculative vision of a future society, it is a description of a mode of human organisation, rooted in the experience of everyday life, which operates side by side with, and in spite of, the dominant authoritarian trends of our society.[16]

His feeling was that we do not have to worry about the boredom of utopia because we will never get there: examples of possibility we can become through a multitude of peaceful, mutually supportive actions and adjustments.

Like Colin, I was drawn to a curiosity more in what 'ordinary' (sic) individuals do. Colin was happy to join me in this project. Those who use allotments express a poetic relation in what they do and in its relationalities with soil, trees, sheds and others, relationally patterning, foremost in a metaphorical sense, the spaces that emerge. Patterning with materiality but also patterning their own lives and in relation with others doing the same, with mutual regard and support. Situated practice and performance builds, reassures and agitates, reciprocity typifies most (if not all) allotment practice. In turn, a curious combination of intense co-operative engagement and the self almost lost in a wider intensity of events enlightens our thinking regarding how landscape occurs: in the practice of those relations.[17] Landscape no longer recedes to the institutionally bespoke; it occurs through our living.

As his philosophy drawn from Kropotkin reasoned, these relations featured great strength from the distinctive combination of mutual aid and self-reliance. Ward pointed out, this relationship is not all altruism. Giving and sharing enable individuals to feel part of the community through their actions, underpinned by a feeling of inclusion in the community, aware of the importance, socially, culturally, economically of one's involvement in its continuity. They feel a stake in the relationship that also may be reciprocated. On this scale, having the right to give, to choose and to exchange freely, form an important bond between people. In ways, in a market where you can give and take, the currency is co-operation. It is especially welcome in communities that feel themselves ineligible for the benefits available to wider society, and thus excluded from relationship with that society. Allotments are often held together in much the same way as other local mutual aid networks such as Local Exchange Trading Systems, credit unions and shared savings groups.[18]

Through that shared work I was able to be, almost every week for over two years, in direct contact with his ideas and his way of working, and finding deeply his utmost sincerity. We kept in touch for years after that. This collaboration or co-operation progressed mutually. Through that time, already in his mid-sixties, he seemed always to be going somewhere – Brighton, Manchester – to give a talk to enthusiastic audiences that usually included many young people.

The town planner Patrick Geddes was one of his main influences.[19] His relationship with town planning bears apparent contradictions, his distaste for the bland ordering of housing as against self-build and his admiration for the UK post-war New Towns, for example. Yet in his BBCTV programme *New Town, Home Town* 1979 he articulates, through the vocal testimony of new town dwellers, the release that these were felt to give amongst the people who moved from unhealthy, cramped and poorly constructed housing; cramped in at least two ways: densely built up with small spaces outside literally to breathe and also cramped in their lack of human opportunity, facilities, spaces of play. His realisation was in both the importance of opportunity to explore one's life and friendships around and the limits to the sociability amongst often too easily romanticised ideas of close communities, amongst squalor and blocked lives. Whilst regarding new towns, at least in their earlier days of facility provision, as full of opportunity for self-creativity, he does not shy away from criticising the debilitating bureaucratisation that crept through in the shaping of people's living space. He valued planning as a means, often more potentially so than actual, to hold the risks of a market free-for-all; to be flexible enough to protect the poor and especially those in need of a home. His interest in human living space and people's relationships with their surroundings was spurred on by his having worked as an architectural assistant during much of the time he edited *Freedom*.

Leisure and other anarchies

Geographies of leisure are rarely thought of in relation to notions of anarchy. When this has been the case, it has been in association with chaos, even neo-liberalism. The writings that flowed from Colin's typewriter were never regarded as chaos, or indeed neo-liberal! Just as his work on *Anarchy* – to which he brought excellent writing and gave a wider media engagement with excellent and inspiring, creative artwork – was far from chaotic.

Colin's anarchy championed human life, supported the post-war development of new towns; he wrote two classics (amongst many others) *The Child in the City*, and *The Child in the Country*. He was deeply aware, having grown up in London's northeast end, of the importance of opportunity particularly amongst the underprivileged. Yet he was never embittered; nor did he feel against anyone, yet he was quick to identify the absurdity of renaming human beings as consumers long before the Economic and Social Research Council major study on consumption. He frowned at the loss of diversity and variety in their associated richness and potential amongst which children could play as 'free space', that dwindled in the name of efficiency, not always so efficient for the development of rich lives amongst the poor. Freedom and creativity in leisure were at the roots of what Colin felt were routes to a *rich* life, the adjective not to be confused with *financial* riches.

Leisure geographies generally have a connection to state intervention and, more recently, commercialisation. Especially earlier recruits came from experience in or with local councils in the UK and their leisure provision and town planning. Now concern mounts in response to austerity measures that close local youth halls, privatise once public spaces such as shopping malls and select public open spaces for commercial development.

Perhaps Colin's support for new towns had stemmed from a hope that they would yield opportunity more diverse, and sites more publicly available for a diversity that often has been lost or was never 'built in'. His support was certainly not for a top down corporation detached from direct local democracy. I often called him, and introduced him as a social anarchist. I feel sure that he was. A strange collaboration occurred when Ward co-authored the book *Sociable Cities*.[20] He wrote this time, surprisingly, with the *Enterprise Zone* and *London 2000* author Peter Hall. *London 2000* epitomised the heavily detached, bureaucratic and statist 1960s planning style; pushing motorways through poor areas of the city, engineered design disrupting the lives of thousands of people not consulted in the process. Ward came from a different source. The book was founded on the idea that had long appealed to Ward, of fresh starts for people's lives through the carefully designed Garden City, yet with spaces 'left open' for imaginative play.

After a handful of attempts over a century ago, the eventual postwar new towns in Britain were much less humanly engaged and engaging. The new book again addressed the possibilities of building new settlements that would have a human grounding. Ward became increasingly ill at ease when the progressive ideas of one of his mentors, Patrick Geddes, turned into a highly bureaucratic machine that organised and built with little attention to human life. Ward's argument was for a new town, or garden city, that provided infrastructure but welcomed self-build, the way Ward saw a progressive future in the freedom of creativity and doing so more affordably.[21] Ward pursued Segal, his mentor and American anarchist thinker and campaigner, in promoting self-built intervention as a decent, accessible creative possibility from improved shanty towns to the western world.

To a degree Ward found an exemplar of self-build-based garden cities in coastal areas of England and especially east of London. These areas, scattered and now in decline, were the Plotlands, built by returning servicemen and others moving out of crowded cities after the First and then Second World War.[22] Buying a cheap plot, on low cost, poor quality land, households were able to build their idea of a home, not reliant upon either local council or capitalist designs (who predominantly built homes paid for by the councils); masterplans or hands-off building. They could build step by step as they liked and could afford, to their choosing: they participated directly in what they were going to live in. Over the decades that followed, some areas may have declined; but many have gradually been enhanced, with refreshed gardens, new building improvements and so on. Alas, they lack the infrastructure that Ward saw as essential. It is no surprise that many households have moved away in the absence of things like proper drainage.[23]

The combination of state support and the freedom *to do* strikes at the core of Ward's adaption of anarchist thinking from Kropotkin to the mid-twentieth-century Paul Goodman. It also exemplifies Ward's more socialist side of concern for the poor. That coupled with his concern for education in living, in engagement with one's local environment, for which he wrote several books and many articles, especially directed towards teachers and their creativity too, informed the power of his books on children's lives in city and country.[24] Moreover, whilst he encouraged opportunity, freedom and creativity to empower poor households to do, themselves, he acknowledged that their energy was not inexhaustible, and so once again required state support. He was concerned for working-class culture; he saw freedom as central to its release from poverty and acknowledged the rich diversity that immigrant people bring to invention in non-predictive freeplay. In similar ways the town or city child was also a country child, desirous of green space too. In this, as in other spaces that Ward examined, there is a yearning for a more participative role for the local council or other responsible body; working with households, providing opportunity and safety assistance but not dominating the scene.

Perhaps at first even more ironically, Ward, with the geographer-planner Denis Hardy, wrote encouragingly about the early years of holiday camps. Therein, they unearthed – amidst their managed components of instructing when campers were to wake up in time for breakfast – a freedom to play, for adults and children alike, with plentiful facilities all around them, however much competitions ordered parts of the day.[25]

His work on the book we shared brought Colin perhaps more directly into anarchic contributions to green ideas and ecological attitudes: an examination, both historical and contemporary, of small plots that people rent for the growing of food, emerging from a need for food for the poor today to be a focus of freedom to grow and how to grow one's own; to have access to land to feed a family; open opportunity to cultivate as one wishes; being creative with materials; recycling and making material and metaphorical landscape in the process. We documented from archives and long informal interviews. These are allotments or community gardens.[26] Moreover, there is a strong social grounding of conviviality in the ways in which plotholders spend their time on their plot. Recycling, typical of numerous allotments, may be more to do with saving money than saving the planet, yet for many it is both. Distinctively, whilst there may be some mutual competition, the character that binds them together socially and culturally is self-help combined with mutual aid: sharing of crops, seeds and tips of cultivation, help in working the ground, a new way of social organisation in the sharing of what is grown, and mutual reliance in the gift relationship amongst them. The allotment site is an alternative to the street that became a channel for cars; it is also an alternative to barren fields, supermarkets and their carparks that speak of airmile fuel. These plots continue to provide seeds of possibility and progress.

All types of community gardens can exhibit particular emphasis on this dimension of plotholding, managing their sites together. Increasing numbers of allotment holders do the same now. From these ventures the community derives strength.

In the posthumous collection of a number of Colin's essays and talks, *Talking Green*, in which he drew threads that had grown through his life, his concern for pollution and green leisure space emerges. He makes a strong case for the distancing of business and science from human beings at large. In one of its written talks, 'The Green Personality', Ward links Kropotkin's *Fields, Factories and Workshops* with the contemporary green personality: 'It pleads for a new economy in the energies used in supplying the needs of human life, since these needs are increasing and the energies are not inexhaustible'.[27]

Around the time we met, Colin had written that in places like allotments and community gardens, there was a mutuality amongst cultivators and other users of these places that offered seeds for possibility and social progress. But for him, allotments provided significant degrees of freedom and creativity too. In this we found great mutuality of ideas and approach. For him, the child in the city and in the country needed leisure space for creativity and experiment.

His *The Child in the City* combined individuals' voices and illustrations by the photographer Ann Golzen that expressed an intensity, variety and ingenuity of urban childhood: not objectifying the human being but letting them speak in an image, certainly not mere 'illustrations'. In many of the images there is a sense of loss or of unattainability as children look out of a torn net curtain to a world of outside freedoms; others make a fire to enjoy the burning, recycle simple cheap material into an imagined toy. Images of children in poor, crowded cities across the world add further impromptu, serendipitous character and insights to the potential of recycling, colonising spaces where people are planned out; at least where the young are unconsidered.

As with all of Ward's work these are not messages of nostalgia, but an alert as to what can be but often isn't: blocked but imagined opportunity, possibility, hope – despite all this, strands of determined invention. In play he found protest and exploration, but also identity, with place as a significant source of what that identity means or feels like. In where and how play is made Ward saw creativity and a practice of what we might call 'worlding'. His influence was felt in the adventure playground movement, propelled by people at large, as it was in the housing self-build movement. Although for some it can deliver significant advantages for those needing to save for a house; the DIY home surely could be in part considered a leisure activity too.

Ward examined the everyday spaces of young people's lives and how, given the opportunity, they can negotiate and re-articulate the various environments they inhabit. In his earlier text on children in the city, the more famous of the two, Colin Ward explored the creativity and uniqueness of children and how they cultivate what he said to me was 'the art of making the city work'. He argued that through play, appropriation and imagination, children can counter adult-based intentions and interpretations of the built environment.

In *The Child in the Country* he unsettled familiar approaches to rural life and its spaces at that time. In the city many children were often hidden, but there was a greater awareness of their circumstance, even existence of contemporary poor, in the postwar decades. It inspired the geographer Chris Philo to call for more attention to be paid to young people as a 'hidden' and marginalised group in society who frequently feel their life alienated in small communities that can suffer from distinct lack of opportunities in play as, later, in work. Ward, however, was keen to stress the individuality of children and their educational needs.

Typically of Colin Ward, his streets of anarchy were not for barricades, but for opportunity and potential. Understanding children, for example, provided a means to go further in working out ways of enabling them to reach those possibilities, showing great awareness of the potentialities of children in their encounters with space. The challenge lies in an approach to enable children through freedom to discover more of their lives, through a freedom that is also made safe. There is also great value in learning from children as from any individuals and collectives seeking to discover, in many cases to repair, their lives; through which better to enable possibility, but not to think in terms of 'achievement', but to have choice of what to do and how to use.

Other kinds of spaces appealed to Ward. It may seem an absurd contradiction to discover that Ward also wrote a book on Chartres Cathedral, seemingly a gem of institutionalism and the hand of a Great Architect; a rather different kind of space. This is a book in Folio Society high quality and lavish with colour illustrations.[28] Yet the relevance of this book to his anarchic thought lay in its focus. The focus is on the workers, the craftsmen and possibly women who built the cathedral. His text concerns the beauty that these people lent to the work; the quirks of individuals that left their own trace in what they made. The gentle anarchy of the builders of the cathedral, not the highly paid designers or financiers, appealed to Colin, who detested tyrannies, as he understood them to be, of visionary thinking, with its potential for coercion and almost inevitable unevenness of power. I think that Colin's sense of gentleness in political thinking-action came to inform a gentle, everyday politics formed significantly of and through living relations and doings, able in the process to make change happen, in the way Colin saw it.

Concluding thoughts

In the 1890s Kropotkin worried over the expanses of empty fields in parts of England at a time of increased mechanisation and imports: 'we have fields, but few men go in... man is conspicuous by his absence from those meadows; he rolls them with a heavy roller in the spring; he spreads manure every three years, then he disappears until the time has come to make the hay'.[29] Plus ça change. In the 1970s the cultivator and writer John Seymour wrote:

> He knows a man who farms 10,000 acres with three men and seasonal contractors, growing barley. Cut that land [exhausted as it is] up into a thousand plots of ten acres each, give it to a family trained to use it, and within ten years the production coming from it would be enormous.... an extremely diverse countryside, orchards, young tree plantations, a myriad of small plots of land growing a multiplicity of different crops, farm animals and hundreds of happy and healthy children.[30]

Marie Louise Berneri, a respected Italian anarchist thinker and practitioner ended her book that documented utopian thought from Plato to Aldous Huxley, *Journey through Utopia* with her typical insight on William Morris's *News from Nowhere*:

utopias where men [sic] were free from both physical and moral compulsion, where they worked not out of necessity or a sense of duty but because they found work a pleasurable activity, where love knew no laws and where everyman was an artist. Utopias have often been plans of societies functioning mechanically, dead structures conceived by economists, politicians and moralists; but they have also been the living dreams of poets.[31]

Echoing Berneri, Ward's writing too concerned freedoms, self and co-operative creativity and expression; new forms of social organisation; mutual aid and reciprocity unhindered by institutional ordering and controls. For him these threads were already to be found, thrusting gently and offering ideas and potentialities for ever wider take up, to be applied, now, in the spaces where we live.

Colin died in 2010 at the age of 85, and survives in his numerous literatures that express those concerns that he felt deeply. He left me with refreshed ideas and a richer approach to research, one that was both more located in the humanities and listening to the subjective feelings about the world than the social sciences: one that has become increasingly prevalent across many disciplines, and notably, geography.

Notes

1 Chris Wilbert and Damien White's recent collection of Colin's work, *Autonomy, Solidarity, Possibility*.
2 Ward and Hardy, *Arcadia for All: The Legacy of a Makeshift Landscape*.
3 Ward, *The Child in the Country*.
4 Ward, *The Child in the Country*; Philo, 'Neglected Rural Geographies: A Review'.
5 Wilbert and White, *Autonomy, Solidarity, Possibility. The Colin Ward Reader*, xxiv.
6 Ward, *Influences: Voices of Creative Dissent*.
7 Ward, *Influences*, 141.
8 Ward, *Anarchy in Action*.
9 Ward, *Influences*, 74–77.
10 Ward, *Influences*, 80.
11 Buber, *I and Thou*, first published in German in 1923 and translated into English in 1937.
12 Crouch, *Flirting with Space: Journeys and Creativity*.
13 Ward, *Influences*, 90.
14 Ward, *Housing: An Anarchist Approach*.
15 Wilbert and White, *Autonomy, Solidarity, Possibility*, xxvi; Massey, *For Space*.
16 Ward, *Anarchy in Action*, 11.
17 Crouch, *Flirting with Space*.
18 Crouch and Ward, *The Allotment: Its Landscape and Culture*.
19 Ward *Influences*.
20 Ward and Hall, *Sociable Cities: The Legacy of Ebenezer Howard*.
21 Ward, *Influences*, 15–35; Wilbert and White *Autonomy, Solidarity, Possibility*, 71–84.
22 Hardy and Ward, Arcadia for All: The Legacy of a Makeshift Landscape.
23 Ward, 'Escaping the City', in Ward *Talking Green*, 116–126.

24 Ward, *Child in the City; The Child in the Country.*
25 Ward and Hardy, *Goodnight Campers!*
26 Crouch and Ward, *The Allotment.*
27 Kropotkin, *Fields, Factories and Workshops*, quoted in Ward, *Talking Green*, 115.
28 Ward, *Chartres: The Making of a Miracle.*
29 Kropotkin, *Fields, Factories and Workshops*, quoted in Crouch and Ward, *The Allotment*, 35.
30 Seymour, *The Fat of the Land*, quoted in Crouch and Ward, *The Allotment*, 35.
31 Berneri, *Journey through Utopia*, 1.

9 Moment, flow, language, non-plan

The unique architecture of insurrection in a Brazilian urban periphery

Rita Velloso[1]

This chapter examines June's 2013 Brazilian urban demonstrations that started in Belo Horizonte's Metropolitan Area, considering them to be a new type of political action, led by citizens who live on the capital cities' poor peripheries, thus allowing the emergence of new recognisable social actors. These uprisings took place in the outskirts of Greater Belo Horizonte's peripheral neighbourhoods, which are connected to the city by federal roads, where local people built barricades to block the access in and out of the capital. Belo Horizonte, in opposition to Rio de Janeiro, was designed to hide its poor inhabitants. If in Rio the limits between middle class neighbourhoods and *favelas* are not that well defined – always mixing people and places – here the spatial segregation is natural to the point that the local government uses the academic jargon 'extensive occupation' to describe our outlying suburbs. The peripheral population's aims through these insurrections were not simply to pose the problem of replacing local government but to regain control of their territory. Their struggle was waged by the transformation of the traditional urban planning logic, centre/power/margin/oppression, that usually defines large metropolitan areas. Emphatically, their claim is to establish new lines of escape from urban poverty while achieving citizenship, beginning to explore new ways of doing politics.

Moment

In early 2013 I was developing a study on the architecture of uprisings and insurgencies based on two distinct moments in urban history: first, the political and even spatial effect in the 1871 Paris commune and its proposal for a self-managed government in France, and second, May 1968 events, also in Paris, which not only were a culmination of social processes that unfolded into new and radical spatial practices but also reverberated into multiple urban environments in other European and American cities.

Although these were such different occasions, both Henri Lefebvre and Guy Debord analysed them more in terms of their similarities rather than their divergences. Both authors describe the Commune and May 1968 as moments of an urban proto-revolution. Lefebvre argues that these political and spatial experiences could be the basis of an entire urban theory, addressing social and spatial *praxis*, spontaneity, creativity and, in Lefebvre's words, 'reconnaissant l'espace social en termes politiques et ne croyant pas qu'un monument puisse être innocent'[2]; Debord, on the other hand reflects upon the Commune's weeks and events in order to establish his peculiar perspective on political interpretation and the fascinating idea of a councilist government of the city, as an alternative to urbanism. As Debord proposes:

> The most revolutionary idea concerning urbanism is not itself urbanistic, technological or aesthetic. It is the project of reconstructing the entire environment in accordance with the needs of the power of workers councils, of the *antistate dictatorship* of the proletariat, of executory dialogue. Such councils can be effective only if they transform existing conditions in their entirety; and they cannot set themselves any lesser task if they wish to be recognised and to recognise themselves in a world of their own making.[3]

I certainly agree with Lefebvre and Debord that such relevant political hypothesis were formulated on those two stages: the 1871 urban *praxis* in relation to the experiences of inhabitants' participation in urban governments, and the 1968 involvement of groups and individuals in decision making processes of megacities. In addition, the hypothesis developed by Lefebvre and Debord could also be aligned with the very anarchist principles of voluntary association, egalitarianism, direct action and radical democracy. In diverse degrees, those two important moments have influenced contemporary urban thought, through particular thinkers, or 'performatively' through the rebellions themselves[4].

Investigating the past topographic memory of cities can prepare us for present times, where such procedures catalyse interpretations of events that bring people together, and at the same time help us understand the centrifugal actions of urban dwellers when sharing places, building networks and disseminating spatial knowledge. This practice results in a research method that explores these circumstances from the inside out, as it aims to understand the history of political riots through its spatial traces, effects and impacts, often camouflaged by urban morphology, plans or designs. The logic of urban historical and economic arrangements only makes sense when the appropriation of space by social actors is taken into account. Besides, analyses of past or current urban uprisings are only possible when combined with representations that have often been reconstructed from fragments of narratives taken from concepts and tools across disciplinary fields. The underlying, driving question behind this research as a whole has been to understand what those singular insurgent spatial practices mean to the peripheral cities in a metropolitan region; to comprehend them in terms of leaving traces in urban dwellers' 'real life'. In other words, during and after those days of insurgency, how should the relation between social actors and urban government be considered?[5]

How have the conflicts with government contributed to how people resituate agendas and negotiations for social justice?[6] How have those momentary ruptures and everyday struggles re-positioned everyday solidarities or mutual aid between neighbours?[7]

Therefore when in June 2013 demonstrations began in Brazil, so many questions were raised that they had to be placed next to those historical experiences of 1871 and 1968 in order to try and build a hypothesis: a theoretical and empirical interrogation that exposed a changing reality.

At first it was largely publicised by mainstream media that the demonstrations lacked focus, that the actions were diffuse and that people on the streets did not really know what they were protesting for or against. Police and government bodies wondered flabbergasted on national news who and where were the leaders of these movements, who could not be reached by media. Squares and streets were occupied for the very first time in decades, with large demonstrations, and roads blocked: the government, policy-makers and the mainstream media were stunned by urban sites transformed into intense political stages, ruptured momentarily for some days or even weeks by collective acts of dissent. Consequently, if there were no readable images of these political subjects to broadcast, no recognisable social actor in charge of these uprisings, no rallies or evidence of any political party, this kind of *political performance* could and would not be caught in the grids of usual interpretation.

In order to comprehend June 2013's momentum, a significant part of the Brazilian left intellectuality tried to recall their own experiences established during the political revolts during the dictatorship of 1960–1980s, refusing to see that there was definitely something new and radically different in these demonstrations. At that point, they made a plea for categories related to them as social subjects that would allow them to recognise the legitimacy of demonstrations to some extent.

At first glance, what happened in June 2013 is difficult to see through or even to comprehend. Undeniably the assemblage of facts presented could no longer be discussed as enshrined in sociological analysis but perhaps as urban studies. Amid this opacity, which does not reduce the huge validity of these facts – quite the opposite – what can be concluded from this experience is the formulation of hypotheses. It is precisely this hypothetical context (perceptual, linguistic and historical hypothesis) in which I write this current narrative: a moment of formulation aimed at understanding the peculiarities of some of Belo Horizonte's political movements.

Firstly it should be noted that what caught the attention of the regional newspapers between June and July 2013 was the fact that something peculiar was happening in these demonstrations, showing a very specific context: the *modus operandi* was the occupation of Belo Horizonte's historical city centre where people appeared to be having a party or celebration. In spite of this, protesters claimed that their number one public enemy was the International Football Federation (FIFA) because of its building projects for the 2013 Confederations Cup in Brazil. What is not only ironic but also beautiful to have witnessed is the fact that tens of thousands of Brazilians were rallying against football – which culturally and symbolically is part of our national pride.

We could say that the *festival*, a collective and celebrative action presented by Henri Lefebvre as symbol of a proto-revolutionary moment, was present in the June demonstrations. At the same time, there was a great lack, because in Belo Horizonte the demonstration route was always the same. Aside from the marches toward the Soccer Arena Mineirão, they were performed in the historical city centre, along with the best-known public spaces of Station Square, Seven Square, Liberty Square and Savassi Square. At first they were not reported by the biggest city newspapers until the moment that the main local TV channel (Globo news) was forced to report in the evening news. However, as news began to emerge, unexpectedly a series of riots took place throughout the metropolitan area. For our research goals, it seemed somehow meaningful to collect the news and then think about that material and those records because, first, there seemed to be a political movement with a significant territorial impact and second, the riots were performed beyond the administrative city limits of Belo Horizonte in cities located on the outskirts of the metropolitan area.

Between 24 June and 2 July it became increasingly clear how distinctive the demonstrations happening in the 11 cities located in the peripheral centrality of the Metropolitan Area were, which blocked state and federal roads that gave access to Belo Horizonte (BH) city centre. These blockages or protests began at dawn, and negotiations between police officers and the inhabitants of these margins lasted usually until 9am when passage was regained. As June progressed, roads were being frequently closed by demonstrators, to the point where roads from eight out of the 34 cities on the outskirts of Greater BH, were simultaneously blocked. What at first appeared to be a number of irrelevant events gained momentum in the media as these blockages of interstate transportation of goods were affecting trade, industry and the economy.

During this moment in time it was impossible to forecast how many people would be participating on a daily basis, as the uprisings that took place in the suburbs of Ribeirão das Neves, Jaboticatubas and Sabará had been very disorganised. If on one day you had 20 people, a couple of sofas, some sticks and some bikes lying on the road, on others there were hundreds of local people as well as truck drivers adhering to the movement as they tried to pass by these areas. This architecture of insurgencies was completely unpredictable for the following weeks and for this to be reported in the news was very special, as Belo Horizonte is a city that traditionally hides its poor population.

Unlike Rio de Janeiro, where the contact between *favelas* and neighbourhoods, people and places happens all the time, in Belo Horizonte social and spatial segregation is naturalised and incorporated not only into governmental speeches as 'extensive occupation' but also in some scientific texts, papers and research with its invariant: 'the urban design of Belo Horizonte and its outlying suburbs'.[8] To some extent it is undeniable that the capital's historic city centre predominates over all these poor neighbourhoods and workers' lives as it has been designed to be a unique centrality; to be inhabited by only bureaucracy and the upper and middle classes. Irrefutably, territories around the state capital were historically constituted to be segregated ones; however, the events of June 2013 shone a new light on this outdated urban configuration and its restraints.

Although Belo Horizonte is a very old functionalist city, planned in the nineteenth century to be triumphantly modern and utilitarian, it is a kind of urban space that rapidly lost its character due to the ideological device that inspired its design, as well as the exhaustion of a peculiar Brazilian cultural process that is constantly adopting international urban models. Aarão Reis, a Brazilian engineer who coordinated Belo Horizonte's urban design team had been Eugène Haussmann's student at the *École Polytechnique* in Paris and, as a result, the city was planned from the very beginning to oppose insurgency – but only for the emulation of the Haussmanian Paris, as, strictly speaking, it was constructed to ward off any possibility of conflict by ignoring it, pushing it spatially aside and not dealing with it. The city would therefore affirm itself as a spatial structure for a new Republican State, translating into geometrical ideals of order and control.[9]

The issue with urban matrix configuration in a capital city is that its spaces should serve as a model for a particular aesthetic experience combined with the expectation of rational knowledge and cosmopolitanism. This ideology of the city conceived a model of urban space that enabled the transfer of both categories of thought as if to provide transpositions for different contexts. It means adaptation to different time frames and historically variable conditions of possibility. In the end, it was a model conceived exactly in opposition to any kind of symbolic appropriation and everyday spatial practices, which are always local and singular.

In Belo Horizonte, the urban planning team applied the so-called Haussmanian rules for Paris almost schematically – an exact boundary within which the governmental apparatus had legal authority was drawn, and this avenue that circumscribed the urban territory was named *Avenida do Contorno* (Contour Avenue).

Despite remaining unequivocally linked to the State and its rationality, the plans for a modern city in the region of Minas Gerais were never fully completed. Capital of a very stratified region, during its implementation Belo Horizonte suffered the Brazilian economic crises and the effects of the 1929 crash. As a result, only a third of its buildings and places were accomplished, with large construction projects being abandoned in the foundation phase. With a low population density, and no dynamic production or work, where people lived without financial, commercial or technical support, could be described as a city that was in fact a large construction site.

Furthermore, the relation between power structures and everyday life makes it essential to conceive and produce specific types of urban spaces in order to achieve the concreteness of domination. The establishment of political technologies by any governmental state involves the control of the urban territory, which takes place, *par excellence*, by land regulation, partitioning and property laws, or, in other words, by the effective regulation of the use of urban places in which people live their everyday lives. Belo Horizonte could here again be considered as an urban experiment: settling down its inhabitants according to their social background without considering merging different social classes; therefore urban zoning was taken very seriously in order to operate and protect bureaucracy and the bourgeoisie. To be labelled a functionalist city was never an uncomfortable title for Belo Horizonte, as official governmental narratives hid their segregationist assumptions behind the argument of an ever-glimpsed national role: the mirage of the internationalised metropolis in the State of Minas Gerais.

Social conflicts should not be expected here! This was an urban society of public servants, government employees and all kinds of families that migrated from the countryside anxious to preserve and continue living their traditional lives. Workers' claims or demands from working class neighbourhoods were not foreseen, not here in a capital city living under the burden of freedom and republican allegories.

Belo Horizonte is often presented as the centre of economic activity in Minas Gerais, structured as a single centrality. Almost everything that happens here converges to the intersection of the main Avenues Afonso Pena and Amazonas, the north/south and east/west axes that exist since the city knows itself as an urban configuration. The image of the city is not only an abstract one, but a powerful geographic constituency that continuously establishes its compelling spatial boundaries.

Flow

In 2012/2013, Brazil was discussing the emergence of a so-called 'new middle class', due to the good economic momentum the country had been experiencing. Surprisingly, it was exactly this group of inhabitants, according to government statistics, who were objecting, arguing their right to move and access the town centre and claim their right to collective consumer goods. It seemed that for the first time, their voice had reached traditional areas in Belo Horizonte.

During those moments of road blockages, it was clear that the metropolis' poor suburban inhabitants were aspiring for the very same issues fought for in other demonstrations in Brazil and abroad; if on one hand the movment was clearly in alignment with Occupy Wall Street, the Arab Spring and the Anonymous movement, on the other hand participants were also asking questions about their basic needs. As residents wrote in a pamphlet: 'How can a neighbourhood full of rich companies harbour so much poverty?'[10] Questioning infrastructure provision is not a trivial nor a political inquiry but a problematising response. These uprisings have touched upon a matter of political praxis, as its strategy to tackle segregation has become clear. The tactics used by the inhabitants of the outskirts of greater Belo Horizonte in June and July

2013 places them as new political subjects as they close access by appropriating spaces of flow. Peripheral regions demonstrate the inability of historic cities to deal with incompleteness, while improvisation shows at the same time the spatial logic of the urban expansion process.

Traditionally, city centre workers live in Greater Belo Horizonte's poorer cities, leading to private and state property investments that have no connections with social actors and urban subjectivities. Theses peripheral cities in Belo Horizonte's metropolitan area were conceived to receive low-income inhabitants, working in steel mills or the mining industry and construction. As a consequence, there are large low-income housing estates mostly built between the 1950s and 1970s that are, nowadays, sorely deteriorated. Beside this, there are no investments in design for public spaces, or improving neighbourhoods concerning the social stratum of their singular inhabitants. Those peripheral suburbs can be considered as islands with poor connection to the inner city, although they undoubtedly express the social heterogeneity of the metropolitan area.

Groups of individuals and residents show how they understand the correlation of forces in insulating and reinventing their social space through conflicts that are brief and more like ephemeral breaks. Above all, these ruptures are decisive in order for the city-metropolis inhabitants and government to understand the relevance of peripheral residents' place in society, and the importance of those neighbourhoods in terms of being configured as *peripheral centralities*.

Transport interruptions and movements expose the architecture of metropolitan riots, which aim not to take any political power. Metropolitan insurrections do not pose the problem of replacing governments; more importantly their struggle is for the transformation of the logic of centre/power/margin/oppression that defines the capital city. One *territory configured as a peripheral centrality* begins to reverse control schemes in the metropolis and operate with another underlying logic, able to establish lines of escape from urban poverty through demonstrations, riots, insurgencies: decentralised and/or polycentric events, their movements are 'building blocks' towards increasing the power of struggles. It becomes clear that the city (as an appropriated space) is included in more than one specific dimension of poverty; workers – the metropolitan precariat whose homes are in the outskirts – are able to articulate the common struggle in contention for mobility and accessibility.

The urban struggles in Brazilian peripheries explain the exhaustion of a functionalist urban design model adopted for decades in this country: the demonstrations make evident how zoning and other urban politics are thought out; those riots and barricades express how segregation is no less than a spatial category closely linked to political processes and ideologies, applied to reinforce deteriorative everyday life conditions.

Centrality does not have an institutional dimension, it defines an urban area of economic and population density, heterogeneity of uses, functional complexity, diverse concentration of commerce and private and public services. If people in those places are asking for the provision of public space, one can conclude that *the periphery protesting is a becoming-centre*. Their power is produced exactly by the perception of its inhabitants that they are trapped in areas without any urban activity build-up that is aimed at establishing them as a centrality. When the perspectives of economic development and new urban design proposals that are distributed around the Metropolitan Region are analysed, one can comprehend that none of them even minimally designs peripheries which provide decent living conditions for the people.

Centre–periphery is a mosaic of neighbourhoods and spatial-temporalities; people coming and going along their extensive itineraries, back and forth from the centre. Their protests refuse an established and traditional order expressed in urban configurations that expect them to have no strong connections with cultural, technological or even economic activities in addition to being 'productive' proletarians. It is required that people living in that peripheral centrality accept what the historical town centre says centralities in peripheries should be. It is as if the periphery were about to grasp some qualities and facilities that only the old centre has and consequently, the centre, as a political and social reference, adopts a position to tell people living in peripheries what they want, how to get there and what is good for them. The periphery is still waiting to be integrated and the centre has still to fulfil its promise.

But the truth is that the periphery needs different rules than those established in the centre. June 2013 seemed therefore to be a kind of redemption of Belo Horizonte's history. Undoubtedly it has been an occasion to discover new significant spaces in the metropolis, spaces that have been defined as specific locations for crowd movements. There have been small disputes, off-centre and polycentric ones, which have redesigned some details of the city and its capillarity. It was clear that significant transformations occurred in those micro-scale territories that were able to reverberate through metropolitan areas.

Roads were simultaneously blocked with sofas and the mayor refused to negotiate with 'people who placed "some" sofas on the road'.[11] There were three different barricades built, by three unrelated neighbourhoods, along the highway that crossed Ribeirão das Neves – where the first protest started in June – and they all had very similar claims. This caused chaos to transport links, the police could not clear the roads and as television broadcasted images from the buses more people turned up to join the riots, almost turning the event into a party.

In order to understand the logic of clusters and networks in these movements, our group of researchers heard from the inhabitants some answers to the questions addressed to government officers who were in possession of the investment budget of their respective municipalities. However, these answers had no power over decision-making processes that could affect the daily lives of those urban residents.

It was astonishing to note how people understood the structural problems of the city, and at the same time had no information about their own neighbourhood in relation to other neighbourhoods. Sabará, for example, presented the same lack of investment as Rio Manso: its residents understood who were the entrepreneurs, investors, state alliances with real estate, but they still lacked the tools to exercise any control over their territories on a local or micro-local scale.

It seems that the act of protesting had been decisive and supported by residents, with many of them aware that they would only gain any visibility due to the interruption of road accesses, as this touched upon a central point in terms of transportation. This impact had been noted by a large portion of the metropolitan population that, to begin with, were supporting all events that happened in June. However, when closures began affecting supply they pulled out, remarking that 'truck drivers in Chile helped topple Allende' and began warning of the risks posed by a shortage of goods.

It is important to note that the road demonstrations never grew in number of people taking part; in fact, the key point was to understand the architecture of metropolitan flows, as well as what is at stake when popular action disrupts the efficiency of people's daily routines and lives in upper class neighbourhoods. Bikers, motorcycle taxis, couriers and suburban residents, who lay beside the motorcycles on the road, participated in these demonstrations, and children complained about how the authorities had money enough to build the soccer team sports centre but no money at all to provide safe points for pedestrians to cross the main roads, a demand that came after 12 years of waiting for a project that was still not finished!

We can only imagine how ephemeral those events and moments were, and, at the same time, how significant. It took a deep intelligence that can only come from a pragmatic claim and a demand for a good collective consumption – such as water supply and the right to transport and security amongst others. Beyond any doubt, it took a political claim to end the invisibility of poor people inhabiting the fringes of the metropolis. Something very powerful happens when roads are blocked – there is a very singular network made of flows interrupted – denoting how the complex space of a metropolitan area could be momentarily reconfigured as a time of proto-urban revolution – in other words, an effective device for interrupting forms of control and reversing resistance networks.

Language

These kinds of protests in Brazilian peripheries have shown a singular form of political action whose strength lies in generic cooperation, networks of solidarity and mutual aid. Demonstrations that block roads are events in which people are beginning to explore new ways of doing politics. That political praxis, once collectively constructed through people's performances, symbolically reconfigures the landscape of large cities, allowing new relations of collective power that will constitute the expression of many subjectivities. The riots were for those dwellers in peripheries the unique dialogical representations produced in their own language games.

Communicative interaction took place as an effect of mutual articulation by the people who organised the protests, which, in Minas Gerais, should only be considered a popular victory and unfortunately not a spectacular one. It is significant that the conflict could have been amplified as part of the urban history of this capital city; not only by making its peripheries visible to the rest of the city but, above all, by empowering protesters who were finally aware of what was happening in their neighbourhood. The news concerning these demonstrations has certainly resulted in an acknowledgement of this, which is, in itself, a form of self-reflection of subjectivities.

Such modulation of political articulation assuredly exposed a particular conflict to a wider audience, extrapolating the geographical reach of the event itself, and indicating how learning and knowledge can be generally taken today as the very definition of social productivity. The action of protest set in motion a unique form of cooperation rooted in the communicative competency of individuals, or, what has been called since Marx, *general intellect* – an entirely implicated form of cooperation in communicative attitudes and the diffuse creativity of human beings.

Precisely who are these political subjectivities built in the periphery? The crowd – a confluence of many – that has nothing to do with the One constituted by the State, but rather re-determines the unit that traditionally defines people. The crowd as a category of production based upon language and constructed from a network of individuals, is a form of political and social existence of the many as many. In other words, it is the mode of being of many singularities that realise the generic power of speaking when it fits them.

Individuals in protests are both a hybrid and juxtaposition and, because of that, are given unlimited potential of their own. Their power comes from their encounter and is prior to any particular thing or shape in what Virno calls 'collective centrifugal'.[12] The crowd is a plurality in the public scene, in collective action; an attention to common issues is an intersection that is not promise but premise: language, intellect and 'the common faculties of humanity' according to Virno.[13] The act of gathering together 'the many' precedes the moment when they come together to perform insurgent practices. Each person in that crowd is there because they share everyday experience. In a word, they have in common a way of trying to experience and confront the world.

Expression is a matter that is configured to give voice and establish language, finding gaps for claims to appear. As any dweller of these regions quickly realised, it is the power of speech that turned this action into something new – there was no need to be politically engaged in a party or union, simply because, at that time, traditional ways of complaint and contestation would not help them reach their goals.[14] In this centripetal arrangement of subjectivities, which operates through knowledge, communication and language, dwellers are no longer passive consumers of information. People create new collective networks of expression as long as they share linguistic and cognitive attitudes. After all, people speak as inhabitants; that is, they express themselves in exactly the opposite way to the professional technique of speech or 'specialised' discourses.

The everyday performances in public arenas is what mobilises the production of real meaning in the possibility of extracting new forms of significance of our cultural world and of discovering new modes of social expression. All communication in everyday life is productive if it is either a sum of resistance born from expressions or if the claim articulates a moment of life as movement, as argued by Toni Negri.[15]

What subjectivity emerges then as political actors of uprisings? A plural subjectivity that replaces the masses and assumes trans-individual dimensions. Political actors are the sum of the resistance of subjectivities that have a generic faculty of speech – the undetermined power of saying. Speech articulation acts to produce new power relations by those taking part in riots and perhaps that is the main force of the crowd; that current urban uprisings are not manifestations of political representation but actions that put in motion a new grammar for political expression – the right to resistance and to struggle for some right that is worthy of being defended.

Non-plan

For those who live in the suburbs and want to participate in a political praxis, the horizon of expression is the production of their everyday life in addition to particular economic configurations. A significant political and spatial practice concerns establishing ways of cooperation, solidarity, mutual aid. The very realm of production implies taking into consideration all life forms developed in the everyday which, at the end, is configured as a constellation of social relations, habits, customs and tactics of surviving.

For all those who are dedicated to thinking about the metropolitan peripheries' regional planning by taking insurgencies as a starting point, a brand new territorial agenda emerges, requiring an effort to handle multi-scalar autonomies and polycentric interrelationships, For the very first time, since the development of the Brazilian periphery during the 1950s, another new form of territory arises; it is a periphery that is a new type of centrality that establishes another urban hierarchy – a singular new urban arrangement that presents a new hierarchy of diverse centralities, each of them producing its own form of expression and discourses.

As moments of uprisings are able to change the course of plans for the metropolis, what is the power of insurgencies in this transformation of urban forms when the crowd becomes a category that thinks of the crisis of the State-form as the effective foundation for a new urban plan? How should one confront the spatial unfolding of demonstrations that results in interruption and disruption of regulatory frameworks of state planning? Compared to planning based on cohesion – which then results in state coercion over the territory – what is the meaning of this kind of spatial appropriation? These are the questions that will outline a conclusion.

An analytical model can find its coherence around a given situation, an urban event that brings together a certain period, specific dynamics, and social actors whose identities and trajectories come to the surface by the action inscribed in networks and practices that constitute a social space. Therefore, the insurgency is a kind of 'counter-use' place that defies urban analysis. Through uprisings, the exhaustion and/or the impossibility of designing as well as predefining uses of spaces within the urban entirety is evident. However, this praxis does not fit in the categories of planning or urbanism.

In June 2013, the voices from the streets rejected the idea of urban planning by unveiling a new terrain for antagonism, which, in Henri Lefebvre's terms, had to be included in the calculation of conditions of possibility to the substantive Urban.[16] Lefebvre is the author that called urban uprising a moment to think of the common and collective externally from State logics. However, uprisings, riots and insurgencies – all those examples of a moment called proto-revolutionary, in Lefebvre's terminology, must be updated in order to describe the metropolitan life experienced by the crowd. In the neoliberal metropolis, an individual is always exposed to the unexpected, unusual and sudden changes, having to remain her/himself flexible to the changing urban experience. The context and experience of the metropolis is, to a large extent, training for precariousness, always requiring urgent adaptation. Here, in this metropolitan area, one simultaneously lives through precariousness and variability, multi-laterally exposed to the world: 'individuals move in a reality always and anyway renewed multiple times'.[17]

More than ever, urban and spatial thought are requiring the critical aim of reinventing democracy. Organisational and institutional forms must be built that can go beyond State logics, for example, a radically new form of democracy in terms of tacit knowledge, beyond the fallacious division between the technocratic employees of the agencies of urban planning and the participation techniques conducted, not uncommonly, by the same agencies. The realistic search for new forms of political action requires us to imagine how to sustain a radical democracy, 'nothing interstitial, marginal or residual, but the concrete appropriation and re-articulation of knowledge/power, something that nowadays is frozen in the administrative apparatus of the state'.[18]

We must think of confrontation between inhabitants and governments through other socio-political dynamics, mostly referred to as micro-politics, than institutional structures. The collective experience should be returned to the centre of the challenge of creating a new institutional logic of society, able to establish a new community based on solidarity and cooperation – an institutional logic able to replace the experience of many as the centre of our social and political practice.

Consequently, how should a peripheral centrality in Brazil be planned? By overcoming all modernist logics of urban policies that have always been associated with mutual innervation between political and economic powers, favouring the richest social strata which in turns results in a built environment strongly influenced by the location of various social groups which historically has strengthened spatial segregation in favour of the elites.

Politics as praxis should be urgently put in place; a field where fights and struggles take place and strategies are performed that result in conflicts around contingent solutions. Thus one may begin to consider the periphery as an object of urban thought open to equal possibilities.

Perhaps taking advantage of a logical disorder that an insurrectional act demands could be a moment to create new communication channels, new forms and modes of interaction, new lines of asymmetric and destabilising forces that allow us to see a demonstration (when the threshold of what is tolerable creates new resistances), but not limited to it. Or possibly to think about how to question the limits of urban plan strategies; in other words, thinking about how to play various tactical games that aim to understand the irreducible multiplicity of these territories through their names: creativity, deprivation, restlessness, destruction, subjection, art and revolt.

In recent decades, urban theories have affirmed that the streets' radical political activity was shut down due to the ubiquity of television and Internet in the domestic sphere of life; the squares would be forever empty, as the street rally no longer made any sense. Nowadays, conversely, one has the answer concerning relations of power established in people's struggles together in urban places.

Insurgency is increasing around the world, requiring people to reinvent democracy while pursuing a coherent anti-capitalist politics. Wherever there is an uprising, there is a street, a square, a road that was profoundly transformed by people's action, in other words, by a radical appropriation of urban spaces by its users 'connected at various levels with the metropolitan and health technical networks, housing, education, communication'.[19]

The poor, the working class and the periphery are now included in communication and in virtual and metropolitan networks in a productive movement that mobilises knowledge – in all forms of life and each individual's world experience – to produce knowledge and power in their life forms. People now understand that micro-politics tactics can produce effects in macro-politics, forcing modification in its strategies. An insurgency, as a contingent political action that is materialised on the unexpected course of events is a public action, a collective performance – and it does not provide a *finished product* but it is, first, a *process*. This is not about making a revolution, or achieving governmental power; instead of configuring a 'state-taking mentality' or re-establishing a political decision-making sphere in political parties, the uprisings are examples of those emancipatory praxes that are certainly closer to autonomy, horizontal decision-making and direct action. This is about defending plural experiences and spatial uses as a potential site for radicalisation and, by extension, it night be claimed that such principles of anarchism animate these forms of resistance.

In examining these particular events in a Brazilian periphery, our ongoing research on the architecture of insurrections allows the conclusion that any urban plan concerned with its concrete outcomes should consider uprisings as an inescapable part of urban foundations. If there is to be any possibility of transforming urban space and life by spatial planning, it is only in envisaging the emergence of an alternative space, which implies considering struggles, riots, or conflicts as concrete attempts to expand and enrich humanity's perceptual capacity to overcome alienation by appropriation of its spaces of life.

Notes

1　I am extremely grateful to Laura Barbi for her insightful comments and for providing invaluable support and encouragement during the drafting of this chapter.
2　Lefebvre, *La Proclamation de la Commune*, 32, 394 ('acknowledging the political nature of the social space and no longer believing that a monument can be neutral').
3　Debord, *Sociedade do espetáculo*, 179, 116. Debord, *Oeuvres*, 179, 842.
4　The following authors and works have been my sources of theoretical dialogues concerning insurgent spatial practices. Souza, White and Springer, *Theories of Resistance: Anarchism, Geography and the Spirit of Revolt*; Sitrin and Azzellini, *They Can't Represent Us: Reinventing Democracy from Greece to Occupy*; Clover, *Riot. Strike. Riot. The New Era of Uprisings*; Squatting Europe Collective, *Squatting in Europe: Radical Spaces, Urban Struggles*; Van der Steen, Bart, Katzeff, Van Hoogenhuijze, Leendert, *The The City is Ours: Squatting and Autonomous Movements in Europe from 1970s to the Present*. Bower, *Architecture and Space Re-imagined: Learning from the Difference, Multiplicity, and Otherness of Development Practice*; Cupers, *Use Matters: An Alternative History of Architecture*; Miessen, *The Nightmare of Participation. Cross Benching Praxis as a Mode of Criticality*; Hou, *Insurgent Public Space: Guerrilla Urbanism and the Remaking of Contemporary Cities*.
5　Souza, 'Cidades Brasileiras, junho de 2013: o(s) sentido(s) da revolta'.
6　Miessen, *The Nightmare of Participation*, 95; Miessen and Basar, *Did Someone Say Participate: An Atlas of Spatial Practice*, 11.
7　Cupers, *Use Matters*, 10; Hou, *Insurgent Public Space*, 7.
8　Monte Mor, 'Extended Urbanization and Settlement Patterns in Brazil: An Environmental Approach' in Brenner (ed.), *Implosions/Explosions: Toward a Study of Planetary Urbanization*, 110.
9　Salgueiro, *Cidades capitais do século XIX: racionalidade, cosmopolitismo e transferência de modelos*, 22.
10　In many demonstrations at blocked roads people would hold hand-written pamphlets and posters expressing their claims.
11　Ribeirão das Neves mayor's interview to a local cable television channel, 28 June, 2013.
12　Virno, *Gramática da multidão. Para uma análise das formas de vida contemporâneas*, 16.
13　Virno, *Gramática da multidão*, 17.
14　Souza, 'Autogestão, autoplanejamento, autonomia: atualidade e dificuldades das práticas espaciais libertárias dos movimentos urbanos', 81.
15　Negri, 'Dispositivo metrópole. A multidão e a metrópole', 207.
16　Lefebvre, *A Revolução Urbana*, 38.
17　Virno, *Gramática da multidão*, 17.
18　Virno, *Gramática da multidão*, 27.
19　Cocco, 'Revolução 2.0: Sul, Sol, Sal', in Cocco and Albagi (eds), *Revolução 2.0 e crise do capitalismo global*, 11.

10 Future (pre-)histories of the state

On anarchy, archaeology, and the decolonial

Anthony Ince and Gerónimo Barrera de la Torre

Geographers and social scientists have long followed Foucault in using the term *archaeology* as a metaphor for the process of uncovering the buried but power-laden layers of knowledges and ideas on which present societies are often unknowingly 'built'. Archaeology as a term implies a sense of lost history rediscovered; a multitude of stories long-forgotten being pieced together in forensic detail through material remnants and their arrangements. In this chapter, we explore less a Foucauldian and more a literal interpretation of archaeology – as an academic discipline with a particular set of ontologies, epistemologies and empirical insights. We read archaeological scholarship through the 'alien' lens of geography, not to crystallise an archaeological gaze that is supposedly better than geographical perspectives but instead to render our hitherto atomised disciplinary debates open to the possibilities that a conversation of the two may be of use to anarchist historical (and contemporary) geographies.

Of particular interest to us is a set of critical literatures in archaeology that can inform geographical understandings of the state and its multiple forms and trajectories. Through a critical discussion of archaeological treatments of the state, and drawing from a radical perspective that brings together Deleuzian philosophy and complexity theory, we develop a non-essentialist, anarchist and decolonial reading that can strengthen existing scholarship on what, elsewhere, we have termed *post-statist geographies.*[1] It is our intention that this chapter will also contribute to future inter-/trans-disciplinary engagements between the two fields more broadly.

The chapter begins with a brief critical discussion of geographical studies on the state, identifying how geographical knowledges are subtly shaped by statist epistemologies, by drawing on previous works that outline our vision for post-statist geographies. Next, a brief discussion of archaeology and its key schools of thought is followed by three key themes in which we seek to draw from a number of emerging strands of contemporary critical archaeology. The first theme concerns the foundations of the state, considering not only its origins but also its institutional structures and relations. In this section, we argue that drawing from archaeology can help to highlight the state's fragility and contingency, and unsettle the perceived certainty of the state as a permanent, natural and universal fixture in society. Second, building on these foundations, we discuss the ontological underpinnings of the state as a colonial and Eurocentric concept, and question the singular notion of *the* state as one of a diversity of polities that have existed in the past or could exist in the future. By decentring the state from our ontologies and narratives of political organisation, we can decolonise the way we think about it and identify alternatives. In the third theme, we discuss the contributions of archaeology to understanding the state as a mode of coercion and domination, as well as a focal point of both pre-emptive and ongoing resistance. In concluding, while recognising potential limitations of archaeological scholarship, we explore how these contributions can signal an important non-essentialist shift in geographical understandings of the state.

Statism and beyond in geography

Despite Agnew's seminal work on the 'territorial trap',[2] in which he criticised scholars for failing to question the solidity of state borders in analyses of international relations, it has taken quite some time for geographers to engage substantially with the structuring role of the state in our thinking. In recent years, geographers have made significant strides in rethinking the state as a complex assemblage of 'prosaic'[3] and 'ordinary'[4] relations, operating not simply through coercive violence[5] but also more subtle mechanisms of ordering, aid, guidance, measurement and smart technologies.[6] These relations regulate and securitise the movement of people, goods and capital at the borders of the state, but they also operate within the micro-spaces of everyday life (e.g. homes, bodies)[7], as well as far into the territories of states elsewhere.[8] As such, the notion of sovereignty[9] – often considered to be a central facet of state modes of power – is increasingly recognised by geographers as profoundly disrupted by the very conduct of states themselves. However, within these debates, definitional issues continue to plague the state and how we experience it empirically.

The growing complexity of many analyses serves to underline the profoundly vague, slippery concept of the state. This is complicated further by the augmented role of global and supra-regional neoliberal institutions and agreements in shaping the parameters of state-scale governance and creating a 'variegated'[10] meshwork of multi-scalar *de facto* regulatory regimes within what are formally understood as *de jure* singular state spaces.[11] In tandem with these uncertainties there has been a growing acknowledgement of the ways in which so-called 'state-centrism' has limited and shaped geographical imaginations.[12] For example, Moisio and Paasi deploy relationality as a notion that can help to overcome the fetishisation of monolithic imaginaries of state sovereignty in geopolitics literatures. For these authors, their priority is to more effectively 'reflect on how the state perpetually regionalises or territorialises the lives of its citizens in state spaces' and recognise how 'state spatial transformation is inescapably connected with certain policy transfers/policy mobilities'.[13] Juliet Fall's powerful critique of the naturalisation of state borders is another example, in which the author dismantles the foundations of economics scholars' conceptions of space-as-container, outlining how '[r]eification, naturalisation, and fetishisation of boundaries happen simultaneously'[14] through discursive and policy constructions of economic and material spaces.

Despite these developments, there continue to be problematic assumptions embedded at the root of geographical treatments of the state. We have discussed these critical issues in depth elsewhere,[15] but a number of central themes stand out. Perhaps the most striking point is that scholars critical of state-centrism have rarely taken their important concerns beyond the realm of critique, remaining within a broadly statist paradigm rather than developing new ways of knowing the world that step outside the state-centric framework they rightly criticise. This, however, is not due to an explicit support for state-building or nationalist efforts; instead, this overall scarcity of conceptual innovation stems, in our view, from a series of unarticulated statist myths, which underpin most geographical (and popular) understandings of the state. These myths of the state discursively render it as natural, efficient, eternal, politically neutral, and the only possible counterbalance to free-market capitalism. As such, this silent statism is a largely unarticulated epistemological 'fix' that undermines and excludes forms of knowledge, and modes of knowledge production, that operate according to logics beyond a Eurocentric statist framework.

In seeking to destabilise, deconstruct and *overcome* this statist paradigm, then, an intellectual project of developing *post-statist* geographies is necessary. Identifying how the logic of statism operates in our structures of knowing is a necessary first step, and anarchism is the central school of thought from which we can draw ideas and inspiration.[16] A particular concern is the positioning of the state as a reference point around which knowledge is constructed. This has a variety of problems, most obviously reinforcing colonial relations of power within and between states, in which a modern statist paradigm – with coercive power operating from a central point of authority – is mobilised as both the assessment method and the ideal-type of any form of organisation.

In this chapter, we are particularly concerned with the way statist knowledge regimes tend to produce strictly delimited temporal and institutional imaginaries of how polities may be organised. Reading geographical questions through contemporary archaeological literatures, we suggest, can add important empirical and conceptual substance to a post-statist project, as well as shedding new light on the geographies of the state more generally. In doing so, we seek to build a framework for understanding social change that decentres the modern, Eurocentric state form and opens up more plural, anarchistic ontologies of social and political organisation.

Archaeology: when spaces and times collide

Archaeology is the study of material artefacts, bodies and structures to analyse and understand past societies. Although archaeology is necessarily linked to the past, it covers the full spectrum of human existence, from the Palaeolithic Era (beginning around 2,500,000 BCE) to the present day. Despite clear overlaps between historical geography and archaeology in terms of sharing some common research questions and methods (e.g. archival research) and research questions, there has been relatively little effort among human geographers to bring the two disciplinary traditions together. This is in contrast to physical geography, which has developed the field of geoarchaeology to integrate the two disciplines around archaeological concerns, although most of this work centres on geomorphological and paleoecological techniques and perspectives, where positivist methodologies prevail. This is similar to landscape archaeology, where Geographic Information Systems (GIS) and geophysics are used to reconstruct past landscapes in order to analyse changing populations, cultures, economic activities and relations of power.

There are, however, some works that integrate a human geography analysis. Lisa Hill has been notable in this regard, arguing that 'there are many commonalities shared by these disciplines'.[17] Hill notes that in the Anglophone world the two fields have shared common intellectual trajectories since the 1950s and 1960s, first embracing empiricism, then positivism, before the gradual emergence of critical and poststructuralist thought from the 1980s onwards.[18] Hill goes so far as to suggest that something akin to what geographers understand as non- or more-than-representational theory is a commonly held viewpoint among archaeologists. Likewise, echoing the geographical ideas of Marston et al.,[19] '[t]he idea that the world is ontologically flat is now old news to many within the archaeological discipline'.[20]

Although the centrality of these alternative ontologies in archaeological literatures may be somewhat overstated by Hill, in recent years two key schools of thought in archaeology have emerged in contrast to the 'processual' or 'evolutionist' orthodoxy, both focusing on contestations and relations of power. Influenced by continental European social theory, 'post-processual' archaeology has heavily criticised the 'scientific' positivism of mainstream approaches. Initially driven by the structuralism of Claude Levi-Strauss, post-processual scholarship has become increasingly influenced (and autocritiqued) by poststructuralist ideas. Embracing the subjectivity embedded in interpretation, post-processual archaeology refuses objectivity and foregrounds a mode of analysis that draws from both materialism and idealism to produce knowledges that are fundamentally oriented towards understanding human agency.[21]

Contemporary materialist, or Marxist, archaeology is also a rejection of the positivist and 'a-political' methods of processual approaches, but draws its inspiration from a historical materialism that foregrounds the analysis of changing relations of production and power over time. Unlike post-processual archaeologists, these scholars follow a dialectical theory of history and understand the agency of peoples and societies to be bound up with dynamic struggles over material and economic relations.[22] What these two schools share, however, is a rejection of positivism and an explicitly politicised conception of archaeology as a mode of social and historical analysis which foregrounds the way past lives, cultures and polities were shaped by often complex and shifting relations of power.

Importantly, both also share a recognition that these histories can play a pivotal role in constructing or critiquing dominant power relations in the present.[23] As we shall see, these critical schools of archaeology are not without some relatively major problems for broader efforts to construct post-statist frameworks, but they do help us to uncover other ways of viewing the state within a much longer timescale and a more heterodox and fine-grained understanding of the constitution of polities. Building on this latter point, we later engage with scholarship drawing from Deleuzian and complexity theory that presents neither an essentialist nor reductionist approach to the archaeological. Moreover, as we will analyse later, an issue that crosses these different perspectives is a common attitude towards truth, which reflects certain forms of understanding in critical archaeologies.[24] The remainder of the chapter explores these possibilities in more depth.

Complexity and evolution: challenging the foundations of the state

Perhaps the most profound difference between geographical and archaeological treatments of the state is the most obvious distinction. Human geographers articulate the state as a given; as a constant (if uneven) presence in geographical studies and debates. The archaeological record, however, demonstrates that the state – indeed *any* formalised hierarchical structure or logic of social organisation – is a relatively new phenomenon. The earliest states[25] only began to emerge patchily (and often initially as cities) as recently as 3,000 BCE,[26] and the modern state studied by geographers has only been the dominant system of organising and managing polities and territories globally since the late colonial period, i.e. for little more than 200 years.[27] When we recognise these facts, two important observations emerge: first that the state is a relatively new addition to human societies; and second, that states have both *beginnings and ends*.

If we explore these in more depth, there is a great deal more to be said. Exactly how states come about is a topic of considerable debate, but a number of key factors commonly influenced this process, especially the emergence of elites, the threat or experience of war, resource conflicts and urbanisation.[28] Contrary to popular accounts and assumptions,[29] population growth has been shown to have relatively little impact on state formation.[30] In many cases, a number of different factors are believed to have contributed simultaneously to state formation, but the emergence of *inequality* is what produced the conditions in which the first states formed. With inequality came the perceived necessity to protect the new hierarchical order and the accumulated influence of elites through the creation of professional standing armies and bureaucratisation of social organisation, often in collaboration with or drawing from religious and spiritual authorities.[31]

Despite the powerful nexus of new social organisation, religious affinities, and coercive power, very few states – be they early or modern – have lasted more than a few hundred years.[32] Crucially, however, the collapse or decline of states does not mean the collapse or decline of the societies from which they emerged.[33] Across archaeology, the notion of 'generations' of states has become a common term referring to the succession of state-building efforts and subsequent collapses within a given region. For example, Rogers' study of several generations of states in eastern Inner Asia (c. 2,000 BCE to the late eighteenth century AD) indicates the contingency and fragility of the state form as only one of many modes of organising the steppe polities during that period. Rogers concluded that one should look at state formation not only as a point of origin, but also

> consider it as a source of constraints and ultimately systems of value that formed the social continuity, discontinuity, and disjunctures integral to the formation of states, [which] does not necessarily imply continuity of economy or cultural practice [but] continuity within the ideological patterns used by elites to establish and legitimate control.[34]

What Rogers[35] and others suggest, then, is that the state is part of a much longer temporal trajectory. Long-term continuity within their respective polities is underpinned not by identifiable, discrete state structures (which regularly come and go) but by much more 'organic' cultural and ideological affinities that are periodically mobilised strategically by emergent elites. State formation is therefore characterised not only by possibility but at least as much by *constraint*, since deeply held norms and values persist or develop independently of different state generations, and aspiring state leaders must shape their own ruling ideologies to fit these much stronger affinities. And then, even if a polity has been 'captured' by the statist logics of these aspiring elites, the state may not survive for long.

In this context, archaeologists have been keenly aware of the sheer diversity of state forms. Rather than identify a singular notion of *the* state, evolutionary and processual archaeological theory identified a whole host of state-related terms to try and gather the huge diversity of social organisation under the umbrella of the state. They refer to 'petty states', 'segmentary states', 'city-states', 'polycentric states', 'statelets', 'peer polities', 'peer statelets', among many others.[36] This well-meaning effort was drawn from an important contribution made by archaeology, namely that what we call states have in most cases throughout human history not been the dominating, territorially contiguous, bureaucratically integrated, militarily singular institutions – characterised by isomorphic polities, bureaucracies, and economies – that we live in today.[37] In fact, most states (especially before European colonialism) were weak, uneven, unstable and heterarchical, often playing a minimal or highly contested role in their subjects' daily lives. By identifying new terms to classify this jumble of institutional relations, archaeologists have sought to better understand the diversity of state forms.

Efforts to develop a typology of state forms, however, have been critiqued by those who view itemised lists of discrete characteristics as actually serving to obscure the true nature of the state as a manifestation of a certain set of social relations. This is an important point, since not only does it parallel important state-theoretic developments in geography,[38] but also because such typologies reify the state as an eternal reference point from which we must define all other societies:

> [T]here is the very real danger that we are trying to 'fit' our archaeological research on past societies into existing evolutionary typologies, rather than find out how far past social forms were similar or different from those known in the ethnographic record.[39]

This attitude being critiqued is precisely the kind of essentialism that we wish to avoid, since is it both empirically incorrect and allows the notion of the state to be weaponised by a linear imaginary of progress from 'savages' or 'primitives' (stateless societies) to 'civilisation' (state societies). Moreover, the fact that archaeology's focus on material remains leads to an overemphasis on sedentary populations means that other forms of social organisation among migratory or nomadic societies are obscured. As a society becomes more complex, so it is implied, the closer it gets to the ideal form of social organisation; that is, the modern Eurocentric state. As González-Ruibal explains, '[t]he archaeological invention of the concept of "Prehistory" in the mid-19th Century... identifies "Prehistory" as time that preceded authentic (state) history',[40] thus implicitly rendering any logic of social organisation preceding the modern state fundamentally *inauthentic*. Although archaeologists rarely integrate it explicitly into their theorising,[41] the statism of the archaeological orthodoxy represents a deeply colonial logic.

Nevertheless, scholars risk falling foul of their own critiques, in trying to read a diverse range of past societies through their modern lens.[42] A related weakness is the archaeological binary that is drawn between egalitarian and complex societies. The former refers to societies in which little or no identifiable authority is wielded by any individual or group over others, whereas the latter refers to societies with two or more social strata. The rationale behind the distinction is understandable, but in practice 'complexity' becomes a code-word for *hierarchy*. Scant attention is paid to the possibility that complexity can be manifested in multiple ways beyond hierarchy and stratification. The outcome of this simplistic binary is that because egalitarianism is perceived as anathema to the state[43] and the state is associated with modern societies, the principle of egalitarian social organisation is also positioned as inherently incompatible with contemporary society. In other words, this archaeological discourse confines egalitarian and non-state logics of organisation to the distant past.

What, then, of the broader contributions of archaeology to understanding the state's foundations? Despite problematic elements (which are critically explored later), two important points can be made. First, in identifying the vast diversity of logics and structures that run across polities, we must recognise the fallacy of seeking to construct an ahistorical notion of a singular, identifiable state. Assigning an eternal 'essence' to what a state is – a set of empirically measurable characteristics – ultimately plays into a deeply problematic colonial and modern discourse of progress. Second, it is equally troublesome to seek to break the notion down into a range of different state forms, since the definition becomes so broad that it loses analytical usefulness. Geographers' definitions of the state tend also to focus on state characteristics,[44] and in this regard the emerging efforts[45] to understand the state and related concepts (e.g. sovereignty, territory) as a set of social relations could be more productive. Likewise, our own efforts to focus not on the state but on stat*ism* – as a set of organisational logics[46] – is, we feel, another way of developing a more 'relational' view of the state.

Following from this, it is important to attune ourselves as geographers to a far longer and more diverse sense of the temporality of human societies. Archaeological methodologies articulate multiple intersecting temporal fields and chronologies – ranging from gross (e.g. ceramic phases, C14 dating), medium-grained (e.g. stratigraphic analysis of floors and buildings) and fine-grained (e.g. texts) – to build a picture of the multiple rhythms and processes cross-cutting a particular place.[47] Perhaps more importantly, whereas geographical imaginaries tend to implicitly understand the state as a constant presence in all societies, building in an archaeological understanding of states as contingent, time-bound and the results of conscious effort by certain groups, serves to destabilise the seemingly eternal temporality of the state. Through this, it may be possible to open up new theoretical and methodological perspectives that put the state in its rightful historical place, not as an end-point or pinnacle, but as one of a multitude of organisational forms and logics that have existed and may exist in the future.

Ontological limits on the conception of the state

As we stress at the beginning of this paper, we consider the shift from an epistemological to an ontological level a significant matter. Since one of our objectives is to decolonise and decentre our way of thinking about the state in geography, consideration of the naturalisation of statist logics is at the forefront of our reflection. We find in archaeology an important ally in this, since it allows us to examine different experiences in social organisation throughout human history and grasp the complexity of these forms and their representation. But also, as we will address, it gives us more data to transcend our exclusive universalities and go beyond our own codes to understand our present.[48]

Clearly, archaeology recognises and explores the blurred lines between civilisation and barbarism, and documents the variety of forms that surpass these concepts. Even so, many archaeological perspectives and their anthropological interpretation are defined by reductionist thinking and typologies. More than that, as a definitional problem, we know that much like geography, the discipline of archaeology originated as part of the expansionist and colonialist politics and discourses of states. But there are alternative proposals that convey a creative way to understand archaeology as distinct from the 'traditional', positivist, official, neo-evolutionary, etc., perspectives. Against this hegemonic archaeological colonialism, we follow Alonso's[49] outline of critical archaeologies. These are archaeologies that are against reductionism, not only of the representation of past societies but also against the narrowing of thought; archaeologies that tend to refuse and confront the reproduction of inequalities and the status quo.

Alonso[50] critiques processual, post-processual and symmetrical archaeologies since from his perspective they do not succeed in transcending the constraints of colonialist thinking. For example, he enquires: 'how does this epistemology work?' He answers that it serves to 'hid[e] power inequalities derived from the privileged research locations from which the archaeological discourse is produced and from which it is demonstrated (to ourselves, society and institutions) the apparent justice and equity of our work as researchers (and, incidentally, to accumulate cultural/academic capital)'.

This anthropological perspective takes a position that radically changes the point of reference; as Marin Jones[51] explains, an archaeology that considers the existence of an 'outside' past, something that needs to be discovered throughout a concrete epistemology, generates an image of that past that allows its political essentialisation and appropriation. This process entails linear progress, which is the fundamental basis of teleological time in official histories.

For the latter, we follow some suggestive ideas about truth and the production of knowledge from an archaeological perspective. Knowledge does not function to reconstruct or 'interpret' but to construct something new from the archaeological record.[52] Alonso draws on a Deleuzian perspective to introduce other ways of engaging with knowledge construction and the understanding of our world. He argues that 'our ways of knowing reality (epistemology) are directly linked to political questions related to what and how reality is constructed and the knowledge that dwells in it (ontology)'. Thence, we have the problem of truth: since all knowledge is ontological, it is derived from the circumstances of every epoch, culture, person, etc. What Alonso[53] makes evident is that archaeology should not look 'for the truth of things, but to understand its articulations, its organisation, limits and ways of construction', in such a way that 'different attitudes towards truth involve different ways to understand a critical archaeology'. So, 'truth is not something "outside" for someone to discover it, but it is constructed'. This refers us back to what the anarchist Gustav Landauer proposed in his reflection on Mauthner's book *Critique of Language*: 'truth is an absolutely negative word, negation in itself, and for that fact is the theme and goal of every science whose hardwearing results are always of a negative nature'.[54] On this point,

Landauer identifies how the construction of knowledge generated by Modern/ Western science reproduces social inequalities, and therefore the urgent need for other ways of thinking.

Therefore truth is an emerging process, and following Alonso, archaeology should not seek to reconstruct the past *but to construct a new past*; and that is our main idea when we entitle this chapter 'future pre-histories' as part of an ongoing process to critically examine the state and the statisms it produces. In this sense, Gallego[55] considers that 'scientific thinking is not to correspond what is seen with what is said, nor to order or systematise what is conceived, but to problematise, to link an ensemble of singularities throughout their differences'. Later, we examine how some proposals have had this effect and help us to move forward in our understanding of the state.

As we have discussed, these critical archaeologies can challenge and usefully inform our conception of state formation from a decolonial perspective, whereby the contingency, variability and discontinuous transformation of social organisation are foregrounded, as well as contesting the foundations of knowledge regarding the state. Elsewhere,[56] we have examined decolonial perspectives in depth, related to our post-statist ideas. However, we engage with this perspective for its focus on decentring dominant world-views (even Western anarchism[57]) and epistemic paradigms acknowledging 'other' knowledges as equally valid and the intersectionality (race, gender, patriarchy, class) that traverses the imposition of modern/Western perspectives. This could be read as a relativist/postmodern analysis, but what we present here is a vindication of a critical, anarchist perspective; an alternative that opposes hierarchical or coercive imposition of a uniform/hegemonic/official way of discovering and understanding reality. Rather than drift into a hegemonic relativism, we follow Adorno[58] when he postulates that relativism is 'the brother of absolutism [and] it approaches a doctrine' – it is, in all, a limitation of thought. Instead, we need to acknowledge subjectivity as a perspective that has a particular localisation from which knowledge is acceded and the world is conceived as a place from which reality is experienced.[59]

Again, however, we are not considering the world as an ensemble of different views where all have found *the* truth or a piece of it. What we stand for is that in the multiplicity of experiences we will find the possibility to join together and cross-reference complementarily different world images and transcend our own limits to understand the complexity and diversity of the world.[60] It is in this space where geographers might usefully draw from the partial, fragmentary experience of working with the archaeological record in seeking not 'whole' truths but cross-fertilising fragments of lived experience to construct new pasts on other, post-statist knowledges, to bring alternative knowledges and imaginaries into view. Thus, returning to Echeverría,[61] to transcend our codes and exclusive universalities, we must maintain that all world-visions are necessarily incomplete and ignorant of many aspects of other realities.[62] Our task is to render this 'un-knowability' visible and explicit, and to bring different situated knowledges and visions into conversation.

Following these perspectives we find archaeologists could help geographers to challenge the actual/official/hegemonic understanding of the state. Moreover, we recognise the necessity of an ontological level of analysis in geography that can articulate the discipline with other disciplines or philosophies, as Deleuzian philosophy might, but also with other geographies from other world-visions altogether. For Alonso[63] 'through Deleuze, philosophy and archaeology can fit together with a politically-aware complexity theory which could allow us to overtake the challenges of scientific reductionism' and also 'works side by side with social movements in a horizontal manner'[64].

This said, we are not proposing engagement with the ideas of Deleuze (and Guattari) to construct a 'Deleuzian geography', since there is a number of important political issues related to Deleuze that would need to be addressed in relation to anarchist and post-statist geographies. Nevertheless, some elements of Alonso's Deleuzean reading of archaeology are fruitful. For example, 'Deleuzian philosophy embraces immanence and rejects transcendence to give account of transformation and the emergence of the novel from efficient causalities and external relations to their terms'.[65] The concept of *immanence* fights the domination of a certain world-vision and helps more open and decentred perspectives to emerge. It also allows the inclusion of complexity, not simply as part of society's linear progress, but in terms of its incommensurable diversity and plurality. Thus, 'archaeology could apprehend this complexity without the fear of losing explicative potential',[66] and incorporate the 'heterogeneity of numerous perspectives about the real'.[67] In the next section, we draw on Pierre Clastres' work to explore how these proposals might relate to the violent power relations involved in statism and state formation.

Contesting statist logics of power

We have argued that an important contribution made by archaeology has been to render the state as only one of many political structures, relations and effects of non-linear societal change. This heterodox understanding of 'the real' allows us to reposition the state and the statist logics on which it is founded as marginalia of a much bigger and more diverse human story. It also calls us to interrogate more closely the circumstances in which states arose. There is little doubt, even in orthodox archaeological literatures, that the authority of states was rooted not in their positive contributions to societies (e.g. in protecting people from a 'savage' life in a 'state of nature') but in the cultural, moral and spiritual codes that elites mobilised and weaponised against their own subjects to claim legitimacy.[68] Likewise, rather than collapsing into an abyss of chaos and self-destruction, periods after the decline of states and empires were in many ways a story of societal resilience; showing the continuity of those underlying norms and affinities as persisting *in spite of* the existence of a state.[69]

However, one of the main challenges in relating geography to archaeology is the interpretation of collected data. Since a colonial perspective reproduces inequalities through the reproduction of actual social schemes projected in the past, the social theory used for interpretation needs to be inherently rebellious against this dominant perspective. As Alonso argues, social theory 'has been chiefly an accomplice of the status quo spreading [a colonial perspective and] categories to all fields with positivism working as a "self-fulfilling prophecy" which makes the world fit its preconceived moulds'.[70]

One work that has the power to creatively problematise these issues is Pierre Clastres' studies on 'societies against the state'.[71] Clastres' ethnographic work demonstrated not simply how stateless societies rejected the structures of the state as a mode of governance but also – and crucially – actively and pre-emptively resisted any incursions of statist foundational logics or rationales. Not only societies *without* the state, then, these were societies *against* the state. Instead of developing his ideas, which we have already analysed elsewhere,[72] we will briefly examine the repercussions for the possibilities of using this work in post-statist geography as a framework for reading archaeology differently and bringing its insights into a post-statist project.

It is significant, as Campagno explains,[73] that very few archaeologists of the Antique period have used Clastres' ideas for their analysis. However, there are several exceptions, including the edited work by the former author. It is clear in these studies that Clastres' reflections allow scholars to think about societies of the distant past in other ways, and to understand the origin and paths of states through a different frame of reference. We will allude to two main ideas which confront the popular misconception of societies without state as lacking something, as incomplete, and leaving political complexity to developed/Western societies.

First, he exposes the inequalities of previous visions by presenting a new problematisation to consider the question of how the state comes into being. What Clastres achieves is to open new questionings; he has addressed a new problem in the definition of the state beyond previous perspectives (i.e. not only to interrogate the origin and form of the state but also to denaturalise its originary myths). These are questions that allow new knowledge to be produced in a way that undermines the centrality of the state as a reference point. Secondly, Clastres presented a new perspective to understand and reflect on the construction of societies through state formation; that is, 'to understand societies "*with* the state" from the perspective of societies "*against* the state", and no more the societies "*without* state" from the view of the state'.[74] It allows us to think of societies against state not as incomplete but radically different.

Considering the evidence of resistance against the state on the basis of Clastres' work, it is possible to critique and negate the 'naturalness' of states, and societies' allegedly inherent desire for them as a kind of predetermined telos. Thus we follow Gledhill[75] when he asserts that 'It would rather be a matter of seeing resistance to state formation as the inherent human tendency, and a transition beyond the absolute rank chiefdom to "the state" based on "permanent coercive power" as a rare event dependent on unusual circumstances'. From the latter, archaeological records and practices can also be used to justify and support counternarratives and resistances in the present.[76]

Lastly, although our main concern is archaeology, it is notable that anthropological frames and social theories often define archaeological interpretation of data. As such, echoing the proposals analysed through the text, we emphasise 'other' experiences and discourses from contemporary peoples who can provide alternative treatments and visions of spatio-temporal and political organisation. From there we can in horizontal dialogue re-read our specific and situated realities across difference. We can turn to other experiences of communalism that fight the state and prefigure new spatio-temporalities.

With this in mind, we finish with two reflections from Indigenous intellectuals. Jaime Martínez Luna,[77] a Zapotecan thinker, asks how the next generation will achieve a continent without borders, without states. He argues that 'we will achieve that, if we reproduce and strengthen our ways of living that are the solutions to State's ubiquity and the private property that it defends, appropriating the planet, the land, which is of every being that inhabits it'. This relates closely to the (Western) anarchist tradition of prefiguration, in which a conscious reworking of social and organisational relations in the here-and-now is what constitutes revolutionary activity; building a new world through everyday actions and interactions. How post-statist thinking (informed by Clastres) could help present social struggles is further indicated by Ailton Krenak,[78] activist of the *Unión de Naciones Indígenas de Brasil*, who affirms that: 'Pierre Clastres [...] concluded that we are societies that naturally organised in a way against the State; there is no ideology in that, we are "against" naturally, like the wind that makes its own path, like the water of a river that makes its own path, we are making our way naturally which does not support that institution as fundamental for our health, education and happiness'. This hints at how we might mobilise anthropological, archaeological and ethnoarchaeological material in concrete struggles over wellbeing and social justice, decentring the state not only from our knowledge systems but also our practical solutions.

Concluding comments: towards a non-essentialist notion of the state in geography

In this chapter, we have analysed and explored the anarchist possibilities embedded in a conversation between archaeology and geography on the subject of the state and its (pre-) histories. Rather than utilising the established anarchist canon, we have drawn primarily from a diverse range of radical, critical and decolonial thinkers to explore these possibilities. In doing so, we have identified three key contributions. First, the relative vastness of the archaeological record can help to render the state not simply a fragile and contested institution – which is already well-documented in geography – but a young, impermanent and time-bound institution that is in fact an *anomaly* rather than the norm when considered in relation to the far longer temporal trajectory of human existence. This disrupts the linear perception of history as a unitary process that moves towards a singular end-point (i.e. the liberal capitalist state).

Second, the contributions of archaeology are not to be taken at face value, and must be problematised through an awareness of the risks of inferring universalisable 'truths' from fragmentary evidence and the situated reference point of Eurocentric modernity. Finally, in our efforts to read geographical debates on the state through a post-statist lens, it is essential to attune ourselves to the voices and lived experiences of those societies and movements that live beyond and against the state. This may potentially include those existing ostensibly 'within' states but organising and collaborating through other logics, platforms and relations.

We are certain that the ideas presented have the capacity to enhance the possibilities for developing anarchist and post-statist geographies – both historical and contemporary – and allow for the inclusion of a fuller spectrum of organisational imaginaries in human experiences, societies and polities. The imperative to cooperatively construct knowledge across and beyond a multitude of reference points – among different societies, cultures, social movements, academic disciplines and beyond – is of particular relevance for mobilising insights for 'real-life' impact. The latter, we believe, will strengthen the interdisciplinarity of geography, but crucially it could also help to *un-discipline* geography in exciting new ways.

Notes

1 See Barrera and Ince, 'Post-statist Epistemologies and the Future of Geographical Knowledge Production, in Lopes de Souza, White and Springer (eds), *Theories of Resistance: Anarchism, Geography and the Spirit of Revolt*; Ince and Barrera, 'For Post-statist Geographies'.
2 Agnew, 'The Territorial Trap: The Geographical Assumptions of International Relations Theory'.
3 Painter, 'Prosaic Geographies of the State'.
4 Staeheli et al., 'Dreaming the Ordinary: Daily Life and the Complex Geographies of Citizenship'.
5 E.g. Boyce, Banister and Slack, 'You and What Army? Violence, the State, and Mexico's War on Drugs'; Martin and Mitchelson, 'Geographies of Detention and Imprisonment: Interrogating Spatial Practices of Confinement, Discipline, Law and State Power'.
6 E.g. Clark, 'The "Life" of the State: Social Reproduction and Geopolitics in Turkey's Kurdish Question'; Jones, Pykett and Whitehead, 'Governing Temptation: Changing Behaviour in an Age of Libertarian Paternalism'; Staeheli et al., 'Dreaming the Ordinary'.
7 E.g. Mountz, 'Human Smuggling, the Transnational Imaginary, and Everyday Geographies of the Nation-state'; Woodward and Bruzzone, 'Touching like a State'.
8 E.g. Silvey, 'Transnational Domestication: State Power and Indonesian Migrant Women in Saudi Arabia'.
9 Mountz, 'Political Geography I: Reconfiguring Geographies of Sovereignty'.
10 Brenner, Peck and Theodore, 'Variegated Neoliberalisation: Geographies, Modalities, Pathways'.
11 Sparke, 'Political Geography: Political Geographies of Globalisation (2) – Governance'.

12 Brenner, 'Beyond State-centrism? Space, Territoriality, and Geographical Scale in Globalisation Studies'; Häkli, 'In the Territory of Knowledge: State-centred Discourses and the Construction of Society'; Marston, 'Space, Culture, State: Uneven Developments in Political Geography'; Moisio and Paasi, 'Beyond State-centricity: Geopolitics of Changing State Spaces'.

13 Moisio and Paasi, 'Beyond State-centricity', 264.

14 Fall, 'Artificial States? On the Enduring Geographical Myth of Natural Borders', 146.

15 Barrera and Ince, 'Post-statist Epistemologies'; Ince and Barrera, 'For Post-statist Geographies'.

16 E.g. Bakunin, *Statism and Anarchy*; Barclay, *People without Government*; Heckert, 'Sexuality as State Form', in Rousselle and Evren (eds) *Post-Anarchism: A Reader*; Landauer, 'Weak Statesmen, Weaker People!' in Kuhn (ed.) *Revolution and Other Writings*.

17 Hill, 'Human Geography and Archaeology: Strange Bedfellows?', 412.

18 Hill, 'Human Geography and Archaeology'.

19 Marston, Jones and Woodward, 'Human Geography without Scale'.

20 Hill, 'Human Geography and Archaeology', 418.

21 E.g. Shackel and Little, 'Post-processual Approaches to Meanings and Uses of Material Culture in Historical Archaeology'; Shanks, 'Post-processual Archaeology and After', in Bentley, Maschner and Chippindale (eds), *Handbook of Archaeological Theories*.

22 E.g. Chapman, *Archaeologies of Complexity*; Wurst, 'The Historical Archaeology of Capitalist Dispossession'.

23 E.g. McGuire, *Archaeology as Political Action*.

24 Alonso, 'Flanqueando el procesualismo y porprocesualismo. Arqueología, teoría de la complejidad y la filosofía de Gilles Deleuze'.

25 Indeed, as we discuss in more detail below, these early states were only states (in the modern sense) if we take a rather broad, unsophisticated and ahistorical definition of the term. Thus, decoupling complex polities from the notion of statehood is an important move in decentring the state from our imaginaries.

26 Yoffee, *Myths of the Archaic State: Evolution of the Earliest Cities, States, and Civilisations*, chapter 3.

27 For a brief overview of colonial state-building and archaeological scholarship, see for e.g. González-Ruibal, 'Postcolonialism and European Archaeology', in Lydon and Rizvi (eds), *Handbook of Postcolonial Archaeology*', 41–43.

28 E.g. Feinman, 'The Emergence of Social Complexity: Why More than Population Size Matters', in Carballo (ed.), *Cooperation and Collective Action: Archaeological Perspectives*; Rogers, 'The Contingencies of State Formation in Eastern Inner Asia'.

29 E.g. Diamond, *El mundo hasta ayer ¿Qué podemos aprender de las sociedades tradicionales?*, 26.

30 Feinman, 'The Emergence of Social Complexity'.

31 Chapman, *Archaeologies of Complexity*.

32 Marcus and Feinman, *Archaic States*.

33 See McAnany and Yoffee, *Questioning Collapse: Human Resilience, Ecological Vulnerability, and the Aftermath of Empire*.

34 Rogers, 'Contingencies of State Formation', 257–258.

35 E.g. Chapman, *Archaeologies of Complexity*; Tilly, 'War Making and State Making as Organised Crime,' in Evans, Rueschemeyer and Skocpol (eds) *Bringing the State Back In*; Yoffee, 'The Power of Infrastructures: A Counternarrative and a Speculation'.

36 See, for e.g. Marcus and Feinman, *Archaic States*, 10.

37 E.g. Marcus and Feinman, *Archaic States*; Rogers, 'Contingencies of State Formation'; Yoffee, 'Infrastructures'; Yoffee, *Myths of the Archaic State*.

38 Painter, 'Prosaic Geographies of the State'.
39 Chapman, *Archaeologies of Complexity*, 196.
40 González-Ruibal, 'Postcolonialism and European Archaeology', 41–42.
41 See, however, Lydon and Rizvi, *Handbook of Postcolonial Archaeology*.
42 In Chapman's case, class struggle.
43 This belief, at least, would find sympathy among anarchists.
44 E.g. Flint and Taylor, *Political Geography*, 137; Jessop, *State Power*; Robinson, 'The Distinction between State and Government'.
45 E.g. Jones, 'State Encounters'; Painter, 'Prosaic Geographies of the State'.
46 Ince and Barrera, 'For Post-statist Geographies'.
47 Marcus and Feinman, *Archaic States*, 12.
48 Echeverría, *Definición de la cultura*.
49 Alonso, 'Flanqueando el procesualismo y porprocesualismo'.
50 Alonso, 'Flanqueando el procesualismo y porprocesualismo', 28.
51 Cited in Alonso, 'Flanqueando el procesualismo y porprocesualismo', 18.
52 Alonso, 'Flanqueando el procesualismo y porprocesualismo', 17.
53 Alonso, 'Flanqueando el procesualismo y porprocesualismo', 16–17.
54 Landauer, *Escepticismo y mística. Aproximaciones a la crítica del lenguaje de Mauthner*, 89.
55 Gallego, 'Deleuze y la filosofía de la ciencia', 72.
56 See Barrera and Ince, 'Post-statist Epistemologies'.
57 The work of Ramnath places an interesting critic about this. Ramnath, *Decolonizing Anarchism: An Antiauthoritarian History of India's Liberation Struggle*.
58 Adorno, *Dialéctica negativa: la jerga de la autenticidad*, 42.
59 Alonso, 'Flanqueando el procesualismo y porprocesualismo', 20.
60 Landauer, *Escepticismo y mística*.
61 Echeverría, *Definición de la cultura*.
62 Santos, *Refundación del Estado en América Latina. Perspectivas desde una epistemología del Sur*.
63 Alonso, 'Flanqueando el procesualismo y porprocesualismo', 14.
64 Alonso, 'Flanqueando el procesualismo y porprocesualismo', 16.
65 Alonso, 'Flanqueando el procesualismo y porprocesualismo', 18.
66 Alonso, 'Flanqueando el procesualismo y porprocesualismo', 28.
67 Alonso, 'Flanqueando el procesualismo y porprocesualismo', 22.
68 E.g. Chapman, *Archaeologies of Complexity*; Yoffee, *Myths of the Archaic State*.
69 E.g. McAnany and Yoffee, *Questioning Collapse*.
70 Alonso, 'Flanqueando el procesualismo y porprocesualismo', 15.
71 Clastres, *Society Against the State: Essays in Political Anthropology*.
72 Barrera and Ince, 'Post-statist Epistemologies'.
73 Campagno, 'Pierre Clastres, las sociedades contra el Estado y el mundo antiguo', in Campagno (ed.), *Pierre Clastres y las sociedades antiguas*.
74 Campagno, 'Pierre Clastres', 21.
75 Gledhill, 'Introduction: The Comparative Analysis of Social and Political Transitions', in Gledhill, Bender and Larsen (eds), *State and Society. The Emergence and Development of Social Hierarchy and Political Centralization*, 9.
76 E.g. McGuire, *Archaeology as Political Action*; Morehart, 'What if the Aztec Empire Never Existed? The Prerequisites of Empire and the Politics of Plausible Alternative Histories'.
77 Martínez, 'Comunalizar la vida toda', 2.
78 Krenak, 'O eterno retorno do encontro'.

11 On 'Other' geographies and anarchisms

Narciso Barrera-Bassols and Gerónimo Barrera de la Torre

Introduction

The aim of this chapter is to draw on spatiotemporal organisation and the production of landscape-territory (moulding, sculpting) that originate from two different perspectives: the Western anarchist perspective, and the perspective that we have called here 'Other' geographies,[1] such as the indigenous perspectives. Taking into account that there is no essential anarchism – that an idealised anarchism has often been contentious with other ways of being libertarian – we go far beyond essentialism to acknowledge that there is an ample avenue to actualise and territorialise the variety of anarchist proposals. Our premise is that there are intersections and converging points of view, as well as discrepancies between the two perspectives. But with a dialogue and critical understanding within them, it may be possible to generate synergies in a common path through the advent of more egalitarian societies. We could consider this task to be one of finding common ground between different perspectives, the 'partial connections', following Strathern's term.[2]

We acknowledge that these 'two' perspectives in fact embrace a great diversity of world-visions, thus we do not try to reduce all this important baggage nor generate another dichotomous distinction between 'anarchist' and 'indigenous' perspectives. We are not defining and limiting the latter but we would like to stress the persistence of idealised visions from both sides. Defining such adherences for peoples' struggles is a reductionist thought and trying to adjust them to a certain definition of anarchism just leads to reducing the latter as an abstract ideology. If we follow the example of the Ejército Zapatista de Liberación Nacional (EZLN) in southern Mexico it is clear that many times people have tried to truncate and define this movement, from a colonialist and arrogant attitude, but as Subcomandante Marcos once mentioned, neo-Zapatismo is more 'an intuition' than a doctrine.[3]

In particular, we are interested in anarchism's blindness to other possible world understandings, and moreover the omission of 'Other' geographies in the growth of anarchist geographies. We envision anarchism not as an established view but as possibilities to overthrow inequalities and generate better conditions for life at the emergence of the Anthropocene.[4] And this is only conceivable if we root out the idea of essence, of an 'anarchist essence', and we look for emerging and localised actualisations of anarchism(s) in the incommensurable variety of worlds.

We will begin by examining the main principles that characterise an anarchist geographical organisation as a particular landscape/territorial arrangement, considering some practical and theoretical elements. In this first section, we highlight some principles of anarchism in its Western tradition and the proposals generated by it. The second section will focus on aspects that generally define the proposals of Latin American indigenous movements, whose actual emergence has made them political movements of strong importance at the present circumstances shaped by the Anthropocene. This will allow us to contrast and also examine some experiences of encounter between these perspectives. Finally, we present what we consider to be some conditions critical for this dialogue to happen, and we discuss a theoretical framework that incorporates/recognises the multiplicity of worlds and anarchisms from a temporal/spatial point of view.

Our main concerns behind this chapter are the different kinds of critical thought from American (but mainly Latin American) indigenous political movements about anarchism seen as just another Western discourse, and because of that, its colonialist position. In this vein, some experiences that arose between anarchist movements and indigenous peoples show the lack of sensibilities and interest for constructing a critical dialogue and how to learn from each other.

As the final chapter of this book, this text does not intend to present a history of the different libertarian and indigenous experiences but to elaborate in the present context common ground from actual life projects and 'past' episodes. Our perspective differs from a teleological account and presents the different spatial-temporal experiences as part of a way to re-elaborate and re-think the possibilities for the coexistence of diverse libertarian projects.

We consider this as a first attempt to bring together distinct experiences and we acknowledge the limitation of our perspective. Constraints are clear when discussing so distinct and diverse world-visions and projects, and establish 'their' principles or tenets. Although, in this lays what we think is the possibility of a better understanding, the possibility to discuss what are the principles or the common ground to engage a dialogue where multiple worlds converge without an overlaying principle. These are the ambiguities and discrepancies that we recognise not as threats but as potentials for conversations between the different landscape/territorial and human–nonhuman relations.

Anarchist landscapes

In this first section, we will analyse the anarchist perspective from what has been called the Western tradition, which generates a series of proposals for spatiotemporal organisation. As Breitbart mentions, the realisation of this libertarian thought/action programme demands totally new geographies.[5]

Firstly, we must acknowledge that anarchism supposes a series of principles, worldviews and also ontologies, because in their interrelation it expresses singular ways of being on Earth. Particularly, in the relationship and conception of nature we find ideas that diverge from the hegemonic Modern thought in the works of Reclus and Kropotkin. Even so, between these two authors there exist some contradictions, but nevertheless we could affirm that some fundamental principles characterise a geographical anarchist organisation at different scales (from specific organisation to a society as a whole).

These new landscapes entail a different relation with the non-human, a sense of nature that is thought not as an object or something external to humans, or even its opposed entity (strengthened by the Modern dichotomy objective–subjective), but one in which humans are an integral part of nature in their own transformation and in the limits of their own existence. Even, in the work of Reclus we can find that non-humans acquire a self-essence, or a spirit. Clark and Martin[6] mention 'while [Reclus'] studies became increasingly scientific, technical, and minutely detailed, he never abandoned the aesthetic, poetic, and even spiritual aspects of his attitude towards nature but rather synthesised these dimensions in his far-ranging, integrative perspective'. This integral perspective, a critical holism, is one of the fundamental aspects that differentiate anarchism from the hegemonic Modern point of view. Likewise, environmental degradation was perceived as a result of an imbalance within human relations, the inequalities and injustice prevailing from the imposition of models which only beneficiate a few. So, 'the domination of nature is a consequence of human domination'[7] and 'the domination of nature will continue as long as humanity remains under the sway of a vast system of social domination'.[8]

As such, these landscapes are shaped by the practical activation of several principles that characterise them but not by a fixed plan that defines their structure. Thus, we ask ourselves, what are these principles that embody sociospatial anarchist organisation? We base this synthetic approach on the perspectives of Kropotkin, Reclus, Voltairine De Cleyre and the studies of Breitbart on the Spanish Civil war experience.

If we can think about one of the central aspects which characterise a libertarian landscape and which has been at the centre of radical geographies, it is *decentralisation*,[9] in its radical meaning, that rejects the concentration of power, the domination of a particular entity and the development of sociospatial structures from hegemonic powers. This reflects one of the priorities of that kind of landscape, but also involves the capacity of people to organise their activities over a cooperative (and empathic) capacity without authoritarian/centralised competitive structures.[10]

Self-sufficiency and *autonomy* are key issues to eliminate super-specialisation and dependency, thus unsustainable scale economies and inequalities of concentration of economic activities could be reduced through economic integration, considering the necessities from, and potentialities of, the environment.

From this perspective, economic integration based on decentralisation needs to *overcome the issue of the urban–rural dichotomy*; thus it is necessary to integrate the different land uses, reducing on the one hand the distance between the space of production and the places of consumption and, on the other, to prevent the concentration of population, considering small and medium cities as most suitable.

The spatial organisation that linked together the previous aspects is *federated communities* in network-like patterns. These networks made up in different levels, from communities to regions, to international scales, function also to eliminate the centralisation of certain activities and the possibility to interchange not only supplies but knowledge, experiences and advocate for a broader conception of the place of peoples in diversity. The present organisation in Rojava, Syria, is an example of the presence of this principle in the practice of democratic confederalism.[11]

The *commons* or communal sense of land and other elements of the landscape are key in the configuration of anarchist spatialities. Property is 'inextricable from the apparatus of the state', and corresponds to a hierarchical appropriation of landscapes but mostly represents the first moment of unevenness. On the contrary, as Springer states 'the commons is (...) the domain of anarchism'.[12]

One of the main principles of anarchy is *liberty*, and even though it is present in many of the aspects we have already mentioned, there is one with an important spatial imprint: *spontaneity/creativity*. From an anarchist perspective, there are no rigid laws or models in the use of land and resources or the localisation of activities; rather, landscapes are products of cooperation and emergent, ever-changing organisations.

Spontaneity and liberty promote *diversity*, as Reclus noted: 'every people gives, so to speak, new clothing to the surrounding nature' and all contribute to the maintenance of the human continuum. Also Voltairine De Cleyre, a Canadian anarchist, argued that a 'wholehearted acceptance of differences is what freedom (i.e. anarchy), rationally intended and consistently practiced, is mainly about'.[13] Besides, she does not conceive anarchism defined from a particular worldview (read modern, Western), asserting that

> it no longer seems necessary to me (...) that one should base his [her] Anarchism on any particular world conception (...) For myself, I believe that all these [proposals] and many more could be advantageously tried in different localities; I would see the instincts and habits of the people express themselves in a free choice in every community; and I am sure that distinct environments would call out distinct adaptations.[14]

The spatial organisation of anarchist geographies (landscapes and territories) is underpinned by the localisation of every activity on the necessities of communities under an ethical framework which considers environmental characteristics, technics, interchanges and cultural singularities. Integration is founded on the diversity of nature, which allows a higher self-sufficiency.

Finally, an ethical framework allowing us to think, act and be on Earth is based on *equality*, since it is not possible to decentralise, promote diversity (and difference) and acknowledge environmental and cultural necessities without eliminating the dominant relation between humans and between humans and non-humans.

With all of the above mentioned, it is now possible to find dynamic, resilient human/non-human connections (an ethos: or an ethical and moral organising principle of the life-world or a sense of dwelling).

What about De Cleyre's proposal about different anarchism(s) founded by different world conceptions or worldviews? We follow the perspective of different authors,[15] trying to elucidate their ideas of 'Other' anarchism(s), considering, as Ramnath[16] argued, that 'Anarchism [is just] one manifestation of a larger family of egalitarian and emancipatory principles'. In the same vein, some voices discuss European anarchism in relation to the existence of non-European anarchisms to decolonise it, or to decentralise anarchism by transcending its European legacy. Clearly, this is an issue that requires a broader discussion and we hope this chapter promotes a continuation of this analysis and the possibility of opening the discussion to a more integral and contextualised understanding.

On 'Other' geographies and libertarian experiences

The experiences of certain peoples in their relationship with land generate singular forms of thinking it, expressing it, using it and acting within it; in short, constructing a constellation of ontologies. These mirror the construction of singular landscapes, which show the reciprocal and inextricable relationship between nature and culture, interweaved through history and in constant change: that is, they constitute 'Other' geographies.

In this section, we will focus on some principles that characterise the experiences and principles of indigenous communities in Latin America. Obviously, there is a great biological and cultural or biocultural diversity, despite showing unity through the practice of certain shared principles. Also, we consider indigenous cultures in Latin America to be an imbrication or syncretism between ancestral cultures and the European culture, rejecting a vision that idealises indigenous peoples as the only keepers of profound knowledges about nature, or as ecologists. After that, in the next section we will examine some encounters between anarchist and indigenous movements and/or the integration of both discourses/practices, all of which produce possible dialogues.

Table 11.1 Indigenous landscapes[17]

Government	There is not a complete abolition of hierarchies but a circulation of authority and the one in 'command' does it, obeying the voice of the whole community (Mandar obedeciendo). For example, among many of the Mesoamerican peoples, decisions are convened in assemblies where all members of the community actively participate. There is a reciprocal relationship and a tendency to avoid the concentration of power. Reciprocity is a key exchange practice that allows for being in common, communal, thus doing communality (common-unity).
Equality	Communities divide government and religious activities (cargos) to avoid the concentration of power and money. Every post requires spending their own money for the whole community in celebrations (fiestas), wherewith some equality is maintained and reinforced. Gender equality is one of the main problems since in most of the indigenous communities patriarchy and machismo prevail.
Liberty/spontaneity	Liberty is understood in terms of the relations among humans and with non-humans as a community (communality). It is the result of a 'healthy' relationship between 'us' and the 'others'. It is a relational liberty among all members of the community (including nonhumans and deities or supra-human powers, forces and substances). Landscapes can be defined as emergent and dynamic spatiotemporal patterns that are not planned as a result of their contingent characteristic; they are socially moulded according to their shifting circumstances.
Diversity	Using and managing different ecological fringes (niches) promotes diversity. Self-sufficiency and sustained use of landscapes and their elements are their principle goals. Landscapes are moulded in patchwork-like mosaic patterns for maintaining the multiple use of communal goods, producing a vast array of different goods in small quantities, and as self-subsistence rationality, avoiding accumulation.
Nature	Reciprocity defines the relations between humans and non-humans; normally, there is no such thing as 'nature'. Other persons that are not humans include not only mountains, clouds, plants or animals, but their spirits, agency and reciprocity among them and within us; this community inhabits the same landscape with similar rights and responsibilities, including spirits and deities. Landscapes are an expression of this symbolism in the particular worldview of a certain people. Balancing these relationships is always necessary for restoring equilibrium from contingency or surprise; as, humans do not possess control over nature (non-humans). Rituals are fundamental to maintaining this secular and sacred balancing reciprocity.
Sacredness/ethics	Reciprocal relations among humans and non-humans eliminate private property. Territory and landscape acquire essential dimensions of identity, production, reproduction and exchange of life in indigenous communities, since they represent the fundamental elements for life and at the same time the limits of existence. Land is the principal symbolic representation in their worldviews and world-lives that expresses the sacred interpretation of their communal places.
Territory/commons	Territory is conceived as a subject (a living being) that includes and embraces humans and non-humans; a wholeness where life is reproduced as an integral part of the universe (cosmos), and where all living persons (humans and non-humans, including spirits) practice communality via reciprocity for the continuing of life for the commons. Territory is not conceived as a property, but as a common space for constructing identity and belonging. There is a rejection of individuality by assuming that 'all of us depend on all of us'. The divide between nature–culture is blurred by acknowledging that humans are part of nature and that non-humans are part of culture, signifying in practice a contextual and ever-changing nature–culture community (culturaleza, in Spanish).

From all the above, we find fundamental to a better understanding and for the possibility of a dialogue that we must be aware of the meanings that every concept has in different contexts. Moreover, we must recognise that some concepts do not exist or have a completely different meaning among other peoples. For example, 'liberty', which is one of the main foundations of the anarchist tradition, is understood by some indigenous peoples as a colonising idea. Liberty brings the possibility to individualise people against communality, advocating competition and free access to property over the world.[18] These concepts are perceived as the language of the conqueror, and so, as we will discuss in the next section, we emphasise the idea of developing a deeper understanding of other peoples' struggles. We propose this, not only as a way to articulate movements but also to expand our conceptions in order to engage in a critical perspective and truly overcome the inherent limits and contradictions of our world.

An example of these shared knowledge/practice principles is what we have examined in the case of the Mesoamerican Chatino's landscape ontology, related to the ways this indigenous people name, signify and think about the different entities of their environment. It also shows how their own worldview (and life-world) is interwoven with the structure and transformations of the landscape. Our participatory work in Chatino territories is an effort to approach what we have called 'Chatino's geography', which has been suppressed and excluded from every 'natural resource' management or agriculture programme and education system. Drawing on this encounter we confront the ignorance of our discipline (and its Western perspective), trying to comprehend alternative human and non-human relationships through Chatino experiences. Besides, we found it necessary to delve into the nuances of Chatino thinking, to discuss and listen about their cosmology, their sacred places and their meanings, and what they value. In order to approach this ontological level, to really move forward in the understanding of the Chatino dialogue with the earth and life concretely manifested through their landscapes, we require a modest/critical attitude and cooperative forms of knowledge production.[19]

On the other hand, Chatinos historically have sought autonomy, and one of their most critical (intellectual) thinkers named Tomás Cruz Lorenzo, brilliantly reached a critical perspective of his own culture, while at the same time drawing a radical picture of their oppression in their position as colonised people. He denounces the epistemic violence and cultural imperialism that Chatino people historically faced:

> it is taught... that our knowledges are ridiculous after 'science', and that the kids who learn this vagueness, abandon and undervalue our rituals, behaviors and knowledge, which now are considered as irrational, superstitious, absurd and false. The truth is now the Western truth, not the Chatino's truth although this truth allowed us to live for centuries.[20]

He did not believe that the Mexican government could resolve their historical affliction but instead the only way to live behind suppression is by self-determination and the construction of a network of Chatino communities based on assemblies and elder councils. Tomás states that:

> I don't ask the State-government to 'rescue' or continue 'rescuing us'; what I'm doing is a call to indigenous persons to make it clear that our culture disappears because of a lack of self-organisation and analysis. As a great sage who contradicts Charles Darwin said, those who survive are not the strongest or more apt, but the best organised.[21]

This means empathy, rather than competiveness.

Another Mexican example is the historical link between the Yaqui people and Magonists, who had developed mutual understanding and helped each other in historical struggles. Before the Mexican Revolution began, this relation had strong ties and Yaquis recognised the 'sacrifices' Magonists made for their own people; thus, even nowadays they still expect that someday 'Capitalism would disappear from the Yaqui region and the red flag of Land and Liberty would have no more enemies'.[22] For their part, Magonists as Livrado Rivera noted thought that, Yaquis should be 'left alone for they could govern themselves as they pleased', and also 'they could contribute with their own intelligence and work for progress and everyone's well-being'.[23] Until today, the Yaqui people continue their struggle for their land and against territorial dispossession.

Here, we offer just a few emancipatory experiences and specific social projects drawn from different peoples of the Earth, but there are many other examples, as the Zapatistas in México, the South Asian communities elucidated by James C. Scott,[24] or the experiences in India, analysed by Ramnath.[25] What is important to underline here is that Indigenous peoples recently emerged as social political actors against Capitalism and dispossession from land and territories. Their struggles are based on the continuity of life on Earth, counteracting the Anthropocene by thought and practice in libertarian and diverse cultural ways, reproducing a pluriverse[26] – instead of the Modern Western notion of a universe – as a sole and unique world to be shared. One planet full of a pluriverse of worlds (worldviews and world lives) reveals these other geographies. In fact, libertarian social movements contradict the same idea of the Anthropocene. Besides that these political actors live in the most diverse biocultural areas of the planet, and acknowledging that they are, in their resistance, halting the perverse processes that are facing us globally, this new geological era is deepening the chances of life extinction on Earth at the hands of only a few transnational corporations and millionaires. This indicates that the other libertarian geographies and their moulded landscapes and territories are not biocultural hotspots, but 'coldspots' reducing fever on our planet. Instead of the notion of the Anthropocene, we are living the Capitalocene,[27] or in one-world thinking,[28] or the Anthropos-non-seen in the one-world thought/praxis.[29] Simplifying one-world thinking reduces the resistance dialogue among libertarian pluriversal experience. One (sick) planet embracing many worlds helps those looking to counteract the Western notion of Anthropocene.

Critical dialogues among the variety of anarchisms

Libertarian ideas as 'concrete pluriversality'

To begin any real dialogue, there is a need to explore deep within us to construct a critical perspective that combines our own and the other ideas. In this vein, we find illustrative reflections from Ramnath,[30] who proclaims that

> we could locate the Western anarchist tradition as one contextually specific manifestation among a larger – indeed global – tradition of anti-authoritarian, egalitarian thought/praxis, of a universal human urge toward emancipation, which also occurs in many other forms in many other contexts. Something else is then the reference point for us, instead of us being the reference point for everything else. This is a deeply decolonising move.

This change in the reference point invites us to reflect on the definition of a social movement as being an anarchist one, or how we critically assess social movements from only one perspective. We consider it necessary to thoroughly explore modernity as a contradictory process and leave behind those aspects contrary to liberty and equality. We have suggested that anarchism represents an alternative to modernity, but if we really want to integrate pluriversal aspirations with concrete praxis, we should transcend the idea that there is only one or a unique point of reference (a one-world view). We advocate then, that in this planet exist a wide variety of worlds, and some of them are sustained by libertarian experiences that are irrigated by some principles shared with Western anarchism. Following Ramnath, we contend that anarchism should be considered as 'one derivation or subset' of a broader tendency, or as an interesting image that she offers when saying that: anarchism is part of 'the Liberty Tree' being 'a great banya, whose branches cross and weave, touching earth in many places to form a horizontal, interconnected grove of new trunks'.[31]

In the contemporary crisis we are facing – that is, in the Anthropocene (or Capitalocene, as we acknowledge) – it becomes necessary to move towards other ontologies, or other experiences of being on Earth. Thus, we could recognise many libertarian ontologies, but as one of the main aspects that we have found some indigenous local movements to miss, a global identity is a critical and strategic element of this. However, this planetary ignorance is now being strongly reversed by global political movements such as Vía Campesina, as just one of the many contemporary examples in resistance that are now looking to the global and not just the local as a space of struggle to overcome the so-called Anthropocene.[32]

Synergies, discrepancies and critical solidarity

Our proposal for an inclusive theoretical framework that recognises the pluriverse of worlds and anarchisms draws in the principles we mentioned in the two previous sections. The main idea is to construct a new 'us', less incomplete and less ignorant. But for this we need to recognise that every worldview is an incomplete one, as our perspective is ignorant of many aspects of the experience of life (incompleteness). We recognise some parallel or analogous issues between the views we present, such as the maintenance and enrichment of diversity and spontaneity in interactions with non-humans, and also in the search of liberty, self-determination or self-sufficiency as part of autonomous and anti-colonial struggles. The relation and conception of nature is another intriguing issue because, on the one hand, indigenous perspectives tend to converge on a rejection of private propriety, market and power, and also in rejection of modern dichotomies (such as the nature/culture divide), allowing a more integrated view of human experience.[33] This coincides with the anarchist perspective since, for example, 'Reclus's view of humanity's place in nature is dialectical, critically holistic, and developmental. In a sense, it might be called an 'emergence' theory, if it is understood that for him humanity is emerging *within* nature rather than '*out of* it'.[34] Additionally, an integrative viewpoint was clear as he recognised that 'just as in society unity is achieved through recognition of diversity, in nature a unifying harmony is attained through diverse often discordant elements'.[35]

All of these aspects have an imprint on spatial organisation and on environmental management, so we consider it relevant that the most diverse and heathiest landscapes are many of those produced, managed and shaped by indigenous peoples and small scale farmers of the world.[36]

As we mention at the beginning of this chapter, we notice that there have been some disruptive experiences between anarchists and indigenous movements that have led to contentious misunderstandings. It seems that there is a belief in a 'superior' form of anarchism (read Western) and as a consequence some experiences indicate that there is an urge to impose 'anarchistic' programmes as 'better' ways to liberate peoples. All of these are criticised, for example, by Aragorn![37], whose main interest is an anarchism that transgresses its European legacy; a non-European anarchism as he calls it. For example, he argues that

> Anarchist criticism is generally more repetitive than it is inspired or influential. Criticism helps us understand the difference between illusion and reality. But the form that anarchist criticism has taken about events in the world is more useful in shaping an understanding of what real anarchists believe than what the world is.[38]

For him 'an indigenous anarchism is an anarchism of place',[39] and 'a category should exist for every self-determined group of people to form their own interpretation of a non-European anarchism (...) [So] it could be carried more "anarchistically" than when safe-guarded by the current group of Cosmopolitan materialists'.[40] For him, some libertarian principles are very valuable, but he finds the European legacy as a heavy burden that leads to a colonising mode of thought.

Because of this, it becomes clear that the very idea is not to impose external experiences, thus the first step is to talk less and do more listening/doing. We find that the principles of the Western anarchist tradition should consider other worldviews not only valid and contemporary, but necessities for the general wellbeing and it should be open to other ontologies for the sake of diversity/equality/difference.

A dialogue should be based on hearing-and-learning/doing from these other experiences so stronger and more global social networks and organisations could be constructed, more than trying to define if particular movements are *really* anarchistic. Decentering our views and our points of reference would make possible a planet shaped by many worlds, a pluriverse based on social projects of 'non-domination [...] and unity-in-diversity in the self-realisation of the whole'.[41] Even so, we find that in some indigenous movements there is also a lack of a wider perspective, such as global identity and unity, and this is where anarchism could contribute strongly to fertilise all those movements.

Some questions arose from the issues analysed before, particularly related to the idea of the Anthropocene and the actual modern civilisation crisis. If there are 'Other' geographies, which respond to alternative forms of relations with nature, is there only one Anthropocene or are there many? If that is so, each one expresses a different speed of transformation or even shows different directions, or different intentions in transforming the face of the Earth, or radically halting the hegemonic one. Thus, we argue that there is no unique way to overcome Anthropocene, but we can draw from alternative rationalities rooted in concrete spatial and temporal cultural contexts. Moving to other ontologies, promoting decentralisation, confederation, diversity and spontaneity could be a possible way to overcome environmental degradation at the local level with a global conscience.

Conclusion

In this chapter we contrast the Western anarchist tradition with the other non-Western perspectives, such as the Chatino perspectives that until recently were distant perspectives and were (and are) distant to each other and without a fertile dialogue among them. Even so, we suggest that many of these perspectives/projects share similar principles, despite being grounded in dissimilar worldviews and life worlds; thus, above all we need to identify possibilities for generating synergies on the premise that in the contemporary moment it is critical to open up our worldview, to hear more and learn from other experiences as a counteraction to current dispossession and also to at least halt the speed and timing of the capitalist Anthropocene.

In the spirit of convergence and dialogue, we find the perspective of post-anarchism compelling, in which a critique of epistemological and ontological essentialism has been undertaken, abandoning uncritical engagements with science.[42] We consider fundamental this openness to 'Other' experiences, so in particular the 'Other' geographies have an equal place in the construction of alternatives.

We would like to conclude by highlighting some words from a dialogue between the Subcomandante Marcos and Antonio, an old indigenous intellectual with whom he shared experience and knowledge. In one of their many encounters, Antonio told Marcos his ideas about the good dreams and the bad ones. Among other things, he said: 'the world in which we live now is not of our own dream, it is other people's dream; but, there are some really good dreams that we forgot until we began to make them true'. And this elder continued, saying

> that there were times when we dream about liberty, and in the meantime, we dream about the 'Other', and then we spoke to her/him finding out that there was not fear in our words, nor fear in our hearing. In our dream we could be standing side by side with the one that was different from us and without having any trouble at all. Thus we could acknowledge that each of us could be what she/he is, what we are, without any confrontation, without any clash, without anyone who would either rule us or make us obey them.[43]

For this dream to become real, what we need is to surpass dichotomies between resistant ways of acting or thinking about the world, and overcome ignorance and incompleteness of the many libertarian ways that humans have historically and culturally shaped, including the 'Other' geographies, their alternative landscapes and integral territories.

In Mexico and in Latin America indigenous movements and indigenous ontologies represent some of the most lively alternatives but have received little attention from geography and their spatiotemporal proposals have been cast aside, even in so-called critical geography. We live in a critical moment that urges a critical understanding and dialogue with these alternative worlds. More than finding solutions through our pre-conceived perspective, we need to focus on constructing problems for concrete contexts and situations within communities.

Notes

1 We name 'Other' geographies to the different spatial-temporal experiences and practices of peoples on Earth, such as indigenous peoples, that manifest the diversity and richness of world-visions and world-life, see for example Barrera-Bassols, et al., 'Geografías y saberes locales sobre paisaje: un giro disciplinario desde la alteridad'.
2 Strathern, *Partial Connections*.

3 Interview with Subcomandante Marcos [Available at https://www.youtube.com/wa tch?v=PDLssf72C3Y&list=PLE8F91BA575051060]
4 We follow Stephen, Crutzen and McNeill in their definition of the Anthropocene, as a new geological epoch where humans have become global geophysical forces so our societies have a significant role in the transformation of Earth, see Stephen, et al., 'The Anthropocene: Are Humans now Overwhelming the Great Forces of Nature?'. However, in this chapter we will discuss the Anthropocene in relation to the Capitalocene and the Anthopos-non-seen that place a distinction where not all humans and peoples have equal impacts on Earth's transformation.
5 Breitbart, *Geografía y anarquismo*, 30.
6 Clark and Martin, *Anarchy, Geography, Modernity: Selected Writings of Elisée Reclus*, 17.
7 Breitbart, *Geografía y anarquismo*, 14.
8 Clark and Martin, *Anarchy, Geography, Modernity*, 29.
9 Springer, *The Anarchist Roots of Geography: Toward Spatial Emancipation*, 160.
10 Breitbart, *Geografía y anarquismo*, 12.
11 Knapp et al., *Revolution in Rojava: Democratic Autonomy and Women's Liberation in Syrian Kurdistan*.
12 Springer, *The Anarchist Roots*, 10.
13 De Cleyre, 'Anarchism'.
14 De Cleyre, 'Anarchism'.
15 Ramnath, *Decolonizing Anarchism: An Antiauthoritarian History of India's Liberation Struggle*; Aragorn!, 'Locating Indigenous Anarchism'; 'A Non-European Anarchism' and 'Toward a Non-European Anarchism or Why a Movement is the Last Thing that People of Color Need', among others.
16 Ramnath, *Decolonizing Anarchism*, 12.
17 See, Barrera-Bassols and Floriani, *Saberes, paisagens, e territórios rurais da America Latina*; Toledo and Barrera-Bassols, *La memoria biocultural. La importancia ecológica de las sabidurías tradicionales*; Martínez Luna, *Eso que llaman comunalidad*; Escobar, *Sentipensar con la tierra. Nuevas lecturas sobre desarrollo, territorio y diferencia*; de la Cadena, 'Indigenous Cosmopolitics in the Andes: Conceptual Reflections beyond "Politics"; and Blaser, *Storytelling Globalization from the Chaco and Beyond*.
18 Martínez, 'Comunalizar la vida toda'.
19 Barrera, *Ontología del paisaje Chatino: el caso de la región de San Juan Lachao, Oaxaca*.
20 Cruz, 'Evitemos que nuestro futuro se nos escape de las ma\nos', 23–24.
21 Cruz, 'Evitemos que nuestro futuro', 33.
22 Torúa, *El magonismo en Sonora (1906–1908). Historia de una persecución*, 78.
23 Torúa, *El magonismo en Sonora*, 79.
24 Scott, *The Art of Not Being Governed: An Anarchist History of Upland Southeast Asia*.
25 Ramnath, *Decolonizing Anarchism*.
26 Blaser, *Storytelling Globalization*.
27 Soper, 'Capitalocene'. Capitalocene refers to the fact that the Anthropocene, the human footprints in this particular epoch are imprints by the capitalist system. That is, not all humans have the same influence and pressure as those who have the political and economic power.
28 Law, 'What's Wrong with a One-world World?'
29 Blaser et al., 'The Anthropocene and the One-world (or the Anthropos-not-seen)'. The idea of 'Anthropos-non-seen' fundamentally denounces the preclusion of 'Other' and possible worlds, as well as the 'Other' geographies, challenging the idea that all humans generate the same effects on Earth and points to the differentiation of those who concentrate power and have the means to impact severely on terrestrial systems.

30 Ramnath, *Decolonizing Anarchism*, 16.
31 Ramnath, *Decolonizing Anarchism*, 18.
32 Desmarais, *La vía campesina. La globalización y el poder del campesinado.*
33 Martínez, *Eso que llaman comunalidad.*
34 Clarke and Martin, *Anarchy, Geography, Modernity*, 21.
35 Clarke and Martin, *Anarchy, Geography, Modernity*, 19.
36 Toledo and Barrera-Bassols, *La memoria biocultural.*
37 Aragorn! 'Locating Indigenous Anarchism'; 'A Non-European Anarchism' and 'Toward a Non-European Anarchism'.
38 Aragorn!, 'Locating Indigenous Anarchism', 24.
39 Aragorn!, 'Locating Indigenous Anarchism', 21.
40 Aragorn!, 'A Non-European Anarchism', 14.
41 Clark and Martin, *Anarchy, Geography, Modernity*, 27.
42 Springer, *The Anarchist Roots*, 49.
43 Subcomandante Marcos, 'Los sueños buenos y malos'.

Bibliography

Adam, P., 'Eloge de Ravachol'. *Entretiens Politiques et Littéraires* (1892): 27–30.

Adamovsky, E., 'Euro-Orientalism and the Making of the Concept of Eastern Europe in France, 1810–1880'. *Journal of Modern History* 77(2005): 591–628. doi:10.1086/497718.

Adams, M., *Kropotkin, Read and the Intellectual History of British Anarchism*. London: Palgrave McMillan, 2015.

Adorno, T.W., *Dialéctica negativa: la jerga de la autenticidad* [Negative dialectics: the jargon of authenticity]. Madrid: Akal, 2005.

Agnew, J., 'The Territorial Trap: The Geographical Assumptions of International Relations Theory'. *Review of International Political Economy* 1, no. 1(1994): 53–80. doi:10.1080/09692299408434268.

Agnew, J., *Globalisation and Sovereignty*. Lanham: Rowman and Littlefield, 2009.

Alavoine-Muller, S., 'Un globe terrestre pour l'Exposition Universelle de 1900. L'utopie géographique d'Élisée Reclus'. *L'Espace géographique* 32, no. 2(2003): 156–170.

Alonso González, P., 'Flanqueando el procesualismo y porprocesualismo. Arqueología, teoría de la complejidad y la filosofía de Gilles Deleuze'. *Complutum* 23, no. 2(2012): 13–32.

Anderson, B., *Imagined Communities: Reflections on the Origin and Spread of Nationalism*. London: Verso, 1991.

Anderson, B., *Under Three Flags: Anarchism and the Anti-Colonial Imagination*. London: Verso, 2007.

Anguelovski, I., 'Environmental Justice'. In *Degrowth: A Vocabulary for a New Era*, edited by G. D'Alisa, F. Demaria, and G. Kallis, 37–40. New York: Routledge, 2015.

Antliff, A., *Anarchist Modernism: Art, Politics, and the First American Avant-garde*. Chicago: University of Chicago Press, 2001.

Aragorn!, 'Locating Indigenous Anarchism', 2005 [Retrieved from http://theanarchis tlibrary.org/library/aragorn-locating-an-indigenous-anarchism].

Aragorn!, 'A Non-European Anarchism', 2007 [Retrieved from http://theanarchistlibra ry.org/library/aragorn-a-non-european-anarchism].

Aragorn!, 'Toward a Non-European Anarchism or Why a Movement is the Last Thing that People of Color Need', 2009 [Retrieved from http://theanarchistlibrary. org/library/aragorn-toward-a-non-european-anarchism-or-why-a-movement-is-the-last-thing-that-people-of-colo].

Arrault, 'La 'référence Reclus'. Pour une relecture des rapports entre Reclus et l'Ecole française de géographie', pp. 1–14. *Colloque 'Elisée Reclus et nos geographies. Texte et Prétextes'*, Lyon, September 2005.

Asara, V., Otero, I.; Demaria, F. and E. Corbera, 'Socially Sustainable Degrowth as a Social-Ecological Transformation: Repoliticizing Sustainability'. *Sustainability Science*, 10(2015): 375–384.

Aubery, P., 'The Anarchism of the Literati of the Symbolist Period'. *The French Review* 42, no. 1(1968): 39–47.

Avakumovic, I. and G. Woodcock, *The Anarchist Prince: A Biographical Study of Peter Kropotkin*. New York: Shocken Books, 1971.

Avrich, P., *Sacco and Vanzetti. The Anarchist Background*, Princeton: Princeton University Press, 1991.

Avrich, P., *Anarchist Voices: An Oral History of Anarchism in America*. Princeton: Princeton University Press, 1995.

Baer, J.A., *Anarchist Immigrants in Spain and Argentina*. Urbana, Springfield and Chicago: University of Illinois Press, 2015.

Bakunin, M., *Correspondance de Michel Bakounine, Lettres à Herzen et à Ogareff (1860–1874), publiées, avec préface et annotation par Michel Dragomanov*. Paris: Perrin, 1896.

Bakunin, M., *Œuvres. Texte établi par James Guillaume, Tome II*. Paris: Stock, 1907.

Bakunin, M., *Statism and Anarchy*. Cambridge: Cambridge University Press, 1994.

Bankowski-Zullig, M., '"Kleinrussischer Separatist" oder "Russischer Nihilist"? M.P. Drahomanov und die Anfange der ukrainischen Emigration in der Schweiz' [Small-Russian separatist or Russian Nihilist? M.P. Drahomanov and the beginning of the Ukrainian migration in Switzerland]. In *Asyl und Aufenthalt: die Schweiz als Zuflucht und Wirkungsstätte von Slaven im 19. und 20. Jahrhundert* [Asylum and residence: Switzerland as refuge and activity centre of the Slaven in the 19th and 20th century], edited by M. Bankowski-Zullig, H. Urech, and P. Brang, 107–138. Basel and Frankfurt: Helbing & Lichtenhahn, 1994.

Bantman, C. and B. Altena, *Reassessing the Transnational Turn: Scales of Analysis in Anarchist and Syndicalist Studies*. New York: Routledge, 2015.

Bantman, C., *The French Anarchists in London, 1880–1914. Exile and Transnationalism in the First Globalization*. Liverpool: Liverpool University Press, 2013.

Barclay, H., *People without Government*, London: Kahn & Averill with Cienfuegos Press, 1982.

Barrera-Bassols, N., Fernández Christlieb, F. and P.S. Urquijo Torres, 'Geografías y saberes locales sobre paisaje: un giro disciplinario desde la alteridad'. Paper presented in the international conference: *'Los Giros de la Geografía Humana: Desafíos y Horizontes'*, Universidad Autónoma Metropolitana, (UAM-I), 26, 27 and 28 November 2008. México.

Barrera-Bassols, N. and N. Floriani, *Saberes, paisagens, e territorios rurais da America Latina* [Knowledge, landscapes and rural territories in Latin America]. Curitiba: UFPR, 2016.

Barrera de la Torre, G., *Ontología del paisaje Chatino: el caso de la región de San Juan Lachao, Oaxaca. Tesis de Maestría en Estudios Regionales*. México: Instituto de Investigaciones Dr. José María Luis Mora, 2015.

Barrera de la Torre, G.and A. Ince, 'Post-statist Epistemologies and the Future of Geographical Knowledge Production'. In *Theories of Resistance: Anarchism, Geography and the Spirit of Revolt*, edited by M. Lopes de Souza, R. White and S. Springer. London: Rowman and Littlefield, 2016.

Barrucand, V., 'Le Rire de Ravachol'. *L'Endehors* 64 (24 July 1892).

Bassin, M., 'Geographical Determinism in Fin-de-siècle Marxism: Georgii Plekhanov and the Environmental Basis of Russian History'. *Annals of the Association of American Geographers* 82, no. 1(1992): 3–22.

Bassin, M., 'Turner, Solov'ev, and the "Frontier Hypothesis": The Nationalist Signification of Open Spaces'. *The Journal of Modern History* 65 no. 3(1993): 473–511.

Baudouin, A., 'Reclus a Colonialist?' *Cybergeo, European Journal of Geography* 26 May 2003 [Retrieved from http://cybergeo.revues.org/4004 on 3 August 2016].

Bayon, D., Flipo, F. and F. Schneider, *La décroissance, 10 questions pour comprendre et en débattre* [Degrowth: Ten questions for understanding and debate]. Paris: La Découverte, 2010.

Benjamin, W., *'O erro do ativismo:' Documentos de barbárie, documentos de cultura* [The mistake of activism: documents of barbarism, documents of culture]. São Paulo: Cultrix, 1986.

Berdoulay, V., *La formation de l'école française de géographie* [The formation of the French school of geography] Paris: CTHS, 1995.

Berghaus, G., *Futurism and Politics. Between Anarchist Rebellion and Fascist Reaction, 1900–1944*. Providence: Berghahn Books, 1996.

Berneri, M.L., *Journey through Utopia*. London: Freedom Press, 1950.

Berry, D. and C. Bantman (eds), *New Perspectives on Anarchism, Labour and Syndicalism. The Individual, the National, and the Transnational*. Cambridge Scholars Publishing: Newcastle upon Tyne, 2010.

Bettini, L., *Bibliografia dell'anarchismo*, 2 vols. Florence: Crescita Politica, 1972–1976.

Bihl, L., 'L'Armée du chahut: les deux Vachalcades de 1896 et 1897'. *Sociétés & Représentations* 27(2001).

Billig, M., *Laughter and Ridicule: Towards a Social Critique of Humour*. London: Sage, 2005.

Billig, M., 'Comic Racism and Violence'. In *Beyond a Joke*, edited by S. Lockyer and M. Pickering, 25–44. London: Palgrave, 2005.

Billig, M., 'Humour and Hatred: The Racist Jokes of the Ku Klux Klan'. *Discourse & Society* 12, no. 3(2001): 267–289.

Blaser, M., *Storytelling Globalization from the Chaco and Beyond*. Durham, NC: Duke University Press, 2010.

Blaser, M., Escobar, A. and M. de la Cadena, (Paper in progress). 'The Anthropocene and the One-world (or the Anthropos-not-seen)'. *An Introduction to the Pluriversal Studies Reader*. Manuscript: 21pp.

Blunt, A. and J. Wills, *Dissident Geographies: An Introduction to Radical Ideas and Practice*. London: Routledge, 2000.

Bookchin, M., 'Deep Ecology, Anarcho syndicalism, and the Future of Anarchist Thought'. In *Deep Ecology & Anarchism. A Polemic*, edited by M. Bookchin, G. Purchase, B. Morris, R. Aitchtey, R. Hart and C. Wilbert, 31–38. London: Freedom Press, 1993.

Bourdieu, P., 'The Social Space and the Genesis of Groups'. *Theory and Society* 14, no. 6(1985): 723–744.

Bower, R., *Architecture and Space Re-imagined: Learning from the Difference, Multiplicity, and Otherness of Development Practice*. London and New York: Routledge, 2016.

Boyce, G.A., Banister, G.M. and J. Slack, 'You and What Army? Violence, the State, and Mexico's War on Drugs'. *Territory, Politics, Governance* 3, no. 4(2015): 446–468. doi:10.1080/21622671.2015.1058723.

Breitbart, M.M. (ed.), *Geografía y anarquismo* [Geography and anarchism]. Barcelona: Oikos-Tau, 1998.

Breitbart, M.M. 'Anarchism/Anarchist Geography'. In *International Encyclopaedia of Human Geography*, edited by R. Kitchin and N. Thrift, 108–115. Oxford: Elsevier, 2009.

Brendell, A., 'The Growing Charm of Dada'. *The New York Review of Books* 63, no. 16(2016): 22–25.

Brenner, N., 'Beyond State-centrism? Space, Territoriality, and Geographical Scale in Globalisation Studies'. *Theory and Society* 28, no. 1(1999): 39–78.

Brenner, N., Peck, J. and N. Theodore, 'Variegated Neoliberalisation: Geographies, Modalities, Pathways'. *Global Networks* 10, no. 2(2010): 182–222.

Breton, A., *Anthology of Black Humor*. San Francisco: City Lights Books, 1997.

Brettell, R., *Pissarro's People*. Munich: DelMonico/Prestel, 2011.

Brigstocke, J., 'Defiant Laughter: Humour and the Aesthetics of Place in Late 19th Century Montmartre'. *Cultural Geographies* 19, no. 2(2012): 217–235. doi:10.1177/1474474011414637.

Brigstocke, J., 'Artistic Parrhesia and the Genealogy of Ethics in Foucault and Benjamin'. *Theory, Culture & Society* 30, no. 1(2013): 57–78. doi:10.1177/0263276412450467.

Brigstocke, J., *The Life of the City: Space, Humour, and the Experience of Truth in Fin-de-siècle Montmartre*. Farnham: Ashgate, 2014.

Bryan, J. and D. Wood, *Weaponizing Maps: Indigenous Peoples and Counterinsurgency in the Americas*. New York, NY: Guilford Publications, 2015.

Buber, M., *I and Thou* (1937). Mansfield Center, CT: Martino, 2010.

Burke, P., 'Context in Context'. *Common Knowledge* 8, no. 1(2002): 152–177.

Buttimer, A., 'Edgar Kant and Heimatkunde: Balto-Skandia and Regional Identity'. In *Geography and National Identity*, edited by D. Hooson, 161–183. Oxford: Blackwell, 1994.

Cahm, C., *Kropotkin and the Rise of Revolutionary Anarchism, 1872–1886*. Cambridge: Cambridge University Press, 1989.

Campagno, M., 'Pierre Clastres, las sociedades contra el Estado y el mundo antiguo' [Pierre Clastres, societies against the state and the ancient world]. In *Pierre Clastres y las sociedades antiguas*, edited by M. Campagno, 7–36. Madrid: Miño y Dávila, 2014.

Candida Smith, R., *Mallarmé's Children: Symbolism and the Renewal of Experience*. Berkeley, CA: University of California Press, 1999.

Capel, H., *Filosofía y Ciencia en la Geografía Contemporanea* [Philosophy and science in contemporary geography], Barcelona: Barcanova, 1981.

Carey, G., '*La Questione Sociale*: An Anarchist Newspaper in Paterson, N.J. (1895–1908)'. In *Italian Americans: New Perspectives in Italian Immigration and Ethnicity*, edited by L. Tomasi. Staten Island, NY: Center for Migration Studies, 1985.

Carrà, C., *La mia vita* [My life]. Rome: Il Cammeo, 1943.

Carrillo Trueba, C., *Pluriverso: Un ensayo sobre el conocimiento indígena contemporáneo* [Pluriverse: an essay on contemporary indigenous knowledge]. Quito: Abya-Yala, 2008.

Carris Cardoso, L., *O lugar da geografia brasileira: a Sociedade de Geografia do Rio de Janeiro entre 1883 e 1945*. São Paulo: Annablume, 2013.

Casanova, J., *Anarchism, the Republic and Civil War in Spain: 1931–1939*. London: Routledge, 2005.

Cassedy, S., 'Nihilism in Russia and as a Russian Export'. In *New Dictionary of the History of Ideas*, edited by M.C. Horowitz. New York: Scribners, 2005.

Castoriadis, C., *The Imaginary Institution of Society.* Cambridge, MA: MIT Press, 1987.

Castro Azevedo, M.H., *Um senhor modernista: biografia de Graça Aranha.* Rio de Janeiro: Academia Brasileira de Letras, 2002.

Cate, P.D., 'The Spirit of Montmartre'. In *The Spirit of Montmartre: Cabarets, Humor and the Avant-Garde, 1875–1905*, edited by P.D. Cate and M. Shaw, 1–94. New Brunswick, NJ: Rutgers University Press, 1996.

Chapman, R., *Archaeologies of Complexity.* London: Routledge, 2003.

Civolani, E., *L'anarchismo dopo La Comune. I casi italiano e spagnolo.* Milan: Franco Angeli, 1981.

Clark, J.H., 'The "Life" of the State: Social Reproduction and Geopolitics in Turkey's Kurdish Question'. *Annals of the Association of American Geographers* 106, no. 5 (2016): 1176–1193. doi:10.1080/24694452.2016.1187061.

Clark, J.P. and C. Martin, *Anarchy, Geography, Modernity: Selected Writings of Elisée Reclus.* Oakland: PM Press, 2013.

Clastres, P., *Society Against the State: Essays in Political Anthropology.* New York: Zone, 1989.

Claval, P., *La pensée géographique* [Geographical thought]. Paris: Société d'Edition, 1972.

Clover, J., *Riot. Strike. Riot. The New Era of Uprisings.* London and New York: Verso Books, 2016.

Cocco, G., 'Espetáculo e imagem na tautologia do capital-atualidade e limites de Guy Debord'. *Revista Lugar Comum. Estudos de Mídia, Cultura e Democracia* 4(1998): 199–209.

Cocco, G., 'A cidade policêntrica e o trabalho da multidão'. *Revista Lugar Comum. Estudos de Mídia, Cultura e Democracia* 9–10(2000): 61–89.

Cocco, G., 'Revolução 2.0: Sul, Sol, Sal' [Revolution 2.0: south, sun and salt]. In *Revolução 2.0 e crise do capitalismo global* [Revolution 2.0 and the crisis of global capitalism], edited by G. Cocco and S. Albagi, 10–26. Rio de Janeiro: Garamond, 2012.

Cohn, J., *Underground Passages. Anarchist Resistance Culture 1948–2011.* Oakland: AK Press, 2014.

Cole, S., 'Dynamite Violence and Literary Culture'. *Modernism/Modernity* 16, no. 2 (2009): 301–329.

Colebrook, C., *Irony.* London: Routledge, 2004.

Comte, A., *A General View of Positivism*, 2nd edn. London: Routledge, 1908.

Confino, D. and D. Rubinstein, 'Kropotkine savant'. *Cahiers du monde russe et soviétique* 33, no. 2 (1992): 243–301.

Craib, R., *The Cry of the Renegade: Politics and Poetry in Interwar Chile.* New York: Oxford University Press, 2016.

Crainz, G., *Padania: Il mondo dei braccianti dall'Ottocento alla fuga dalle campagne.* Rome: Donzelli, 1994.

Crimethinc Ex-Workers' Collective, *Recipes for Disaster. An Anarchist Cookbook a Moveable Feast.* Olympia, 2005.

Critchley, S., *On Humour.* London: Routledge, 2001.

Critchley, S., *Infinitely Demanding: Ethics of Commitment, Politics of Resistance.* London: Verso, 2007.

Crouch, D., *Flirting with Space: Journeys and Creativity.* Ashgate: Farnham, 2010.

Crouch, D. and C. Ward, *The Allotment: Its Landscape and Culture.* Nottingham: Five Leaves Books, 1997.

Crowder, G., *Classical Anarchism*. Oxford: Clarendon Press, 1991.

Cruz Lorenzo, T., 'Evitemos que nuestro futuro se nos escape de las manos'. *Medio Milenio*5(1989): 23–34.

Cupers, K., *Use Matters: An Alternative History of Architecture*. London and New York: Routledge, 2013.

Dahlmann, D., *Land und Freiheit. Machnovščina und Zapatismo als Beispiele agrarrevolutionärer Bewegungen*. Stuttgart: Steiner-Verlag-Wiesbaden, 1986.

D'Alisa, G. and G. Kallis, 'Post-normal Science'. In *Degrowth: A Vocabulary for a New Era*, edited by G. D'Alisa, F. Demaria, and G. Kallis, 185–188. New York: Routledge, 2015.

De Cleyre, V., 'Anarchism' (1901) [Retrieved from http://www.panarchy.org/indexes/anarchy.html on 14 September 2016].

De Laforcade, G., 'Straddling the Nation and the Working World: Anarchism and Syndicalism on the Docks and Rivers of Argentina'. In *Anarchism and Syndicalism in the Colonial and Postcolonial World, 1870–1940. The Praxis of National Liberation, Internationalism, and Social Revolution*, edited by S. Hirsch and L. van der Walt, 321–362. Leiden: Brill, 2014.

Delgado de Carvalho, C.M., *Metodologia do ensino geográfico (Introdução aos estudos da geografia moderna)*. Rio de Janeiro: Francisco Alves, 1925.

De Maria, C. (ed.), *Andrea Costa e il governo della cittá. L'esperienza amministrativa di Imola e il municipalismo popolare (1881–1914)*. Bologna: Diabasis, 2010.

De Rivero, O. *The Myth of Development: The Non-Viable Economies of the 21st Century*. London: Zed Books, 2001.

Debord, G., *Oeuvres* [Works]. Paris: Gallimard, 2006.

Debord, G., *Sociedade do espetáculo* [Society of spectacle]. Rio de Janeiro: Contraponto, 2007.

De la Cadena, M., 'Indigenous Cosmopolitics in the Andes: Conceptual Reflections beyond "Politics"', *Cultural Anthropology* 25, no. 2(2010): 334–370.

Demaria, F., Schneider, F., Sekulova, F. and J. Martínez-Alier, 'What is Degrowth? From an Activist Slogan to a Social Movement'. *Environmental Values* 22(2013): 191–215.

Deprest, F., *Élisée Reclus et l'Algérie colonisée*. Paris: Belin, 2012.

Deriu, M., 'Democracies with a Future: Degrowth and the Democratic Transition'. *Futures* 44(2012): 553–561.

Desmarais, A.A., *La vía campesina. La globalización y el poder del campesinado*. [Peasants' Way: globalisation and peasants' power]. Madrid: Editorial Popular, 2007.

Di Paola, P., *The Knights Errant of Anarchy, London and the Italian Anarchist Diaspora (1880–1917)*, Liverpool: Liverpool University Press, 2013.

Diamond, J., *El mundo hasta ayer ¿Qué podemos aprender de las sociedades tradicionales?* [The world until yesterday: What can we learn from traditional societies?]. Querétaro: Debate, 2013.

Dogliani, P., *Un laboratorio di socialismo municipalismo. La Francia (1870–1920)* [A laboratory for municipal socialism: France (1870–1920)]. Milan: Franco Angeli, 1992.

Dragomanov, M., 'Les paysans Russo-ukrainiens sous les libéraux Hongrois'. *Le Travailleur* 1(1877): 12–14.

Dragomanov, M., *Le tyrannicide en Russie et l'œuvre de l'Europe occidentale* [Tyrannicide in Russia and the work of Western Europe]. Geneva: Imprimerie du Rabotnik, 1881.

Drahomanov, M., *A Symposium and Selected Writings*. New York: Ukrainian Academy of Arts and Sciences, 1952.

Driver, F., *Geography Militant: Cultures of Exploration and Empire*. Oxford: Blackwell, 2001.

Dugatkin, L., *The Prince of Evolution. Peter Kropotkin's Adventures in Science and Politics*, Charleston, SC: Createspace, 2011.

Dunbar, G., *Elisée Reclus. Historian of Nature*. Hamden: Archon Books, 1978.

Dupuis-Déri, F., 'Is the State Part of the Matrix of Domination and Intersectionality? An Anarchist Inquiry', *Anarchist Studies* 24, no. 1(2016): 36–61.

Eagleton, T., 'The Revolutionary. Is Marx Still Relevant?' *Harper's Magazine*, April 2013.

Ealham, C., 'An Imagined Geography: Ideology, Urban Space and Protest in the Centre of Barcelona's "Chinatown", 1835–1936'. *International Review of Social History* 50, no. 3(2005): 373–397. doi:10.1017/s0020859005002154.

Ealham, C., 'The Myth of the "Maddened Crowd": Class, Culture and Space in the Revolutionary Urbanist Project in Barcelona, 1936–1937'. In *The Splintering of Spain: Cultural History and the Spanish Civil War, 1936–1939*, edited by C. Ealham and M. Richards, 111–132. Cambridge: Cambridge University Press, 2005.

Ealham, C., *Class Culture and Conflict in Barcelona 1898–1937*. London: Routledge, 2005.

Echeverría, B., *Definición de la cultura* [Definition of culture]. México: Itaca y Fondo de Cultura Económica, 2010.

Ehrlich, P.R., *The Population Bomb*. New York: Sierra Club/Ballantine Books, 1968.

Eltzbacher, P., *Der Anarchismus* [Anarchism]. Berlin, 1900.

Escobar, A., 'Degrowth, Postdevelopment, and Transition: A Preliminary Conversation'. *Sustainability Science* 10(2015): 451–462. doi:10.1007/s11625–11015–0297–0295.

Escobar, A., *Sentipensar con la tierra. Nuevas lecturas sobre desarrollo, territorio y diferencia* [Feeling and thinking the land. New readings of development, territory and Difference]. Medellín: Ediciones UNAULA, 2014.

Esenwein, G.R., *Anarchist Ideology and the Working-Class Movement in Spain, 1868–1898*. Berkeley: University of California Press, 1989.

Fall, J.J. 'Artificial States? On the Enduring Geographical Myth of Natural Borders'. *Political Geography* 29 no. 3(2010): 140–147. doi:10.1016/j.polgeo.2010.02.007.

Faudemay, A., 'L'humour et l'anarchisme: quelques indices d'une convergence possible'. *Revue d'histoire littéraire de la France* 99, no. 3(1999): 467–484.

Febvre, L., *La terre et l'évolution humaine*. Paris: Michel, 1922.

Feinman, G., 'The Emergence of Social Complexity: Why More than Population Size Matters'. In *Cooperation and Collective Action: Archaeological Perspectives*, edited by D.M. Carballo, 35–56. Boulder: University Press of Colorado, 2013.

Fellner, G. (ed.), *Life of an Anarchist: the Alexander Berkman Reader*. New York: Seven Stories Press, 2005.

Ferguson, K.E., *Emma Goldman. Political Thinking in the Streets*. Lanham: Rowman and Littlefield, 2011.

Ferretti, F., 'Traduire Reclus: l'Italie écrite par Attilio Brunialti'. *Cybergeo, European Journal of Geography* (2009) [Retrieved from http://www.cybergeo.eu/index22544. html] doi:10.4000/cybergeo.22544.

Ferretti, F., *Anarchici ed editori: Reti scientifiche, editoriali e lotte culturali attorno alla Nuova Geografia Universale di Élisée Reclus (1876–1896)* [Anarchists and publishers: scientifc networks and cultural struggles around Elisée Reclus's New Universal Geography]. Milan: Zero in Condotta, 2011.

Ferretti, F., *L'Occidente di Élisée Reclus. L'invenzione dell'Europa nella Nouvelle Géographie Universelle, 1876–1894*. PhD thesis, Bologna and Paris, 2011.

Ferretti, F., 'The Correspondence between Élisée Reclus and Pëtr Kropotkin as a Source for the History of Geography'. *Journal of Historical Geography* 37(2011): 216–222. doi:10.1016/j.jhg.2010.10.001.

Ferretti, F., 'Républicanisme, migrations et mélanges en Amérique du Sud dans la géographie d'Élisée Reclus (1865–1905)', in *Ils ont fait les Amériques*, edited by L. Dornel, M. Guicharnaud-Tollis, M. Parsons and J.-Y. Puyo, 319–332. Bordeaux: Presses Universitaires de Bordeaux, 2012.

Ferretti, F., '"They Have the Right to Throw Us Out": Élisée Reclus' New Universal Geography', *Antipode*45, no. 5(2013): 1337–1355.

Ferretti, F., 'De l'empathie en géographie et d'un réseau de géographes. La Chine vue par Léon Metchnikoff, Élisée Reclus et François Turrettini'. *Cybergeo. European Journal of Geography* [Retrieved from http://cybergeo.revues.org/26127] doi:10.4000/cybergeo.26127.

Ferretti, F., 'Géographie, éducation libertaire et établissement de l'école publique entre le19ᵉ et le 20ᵉ siècle: Quelques repères pour une recherche'. *Cartable de Clio, Revue suisse sur les Didactiques de l'Histoire* 13(2013): 187–199.

Ferretti, F., *Élisée Reclus: pour une géographie nouvelle*. Paris: CTHS, 2014.

Ferretti, F., 'Inventing Italy. Geography, Risorgimento and National Imagination: The International Circulation of Geographical Knowledge in the 19th Century'. *The Geographical Journal* 180(2014): 402–413. doi:10.1111/geoj.12068.

Ferretti, F., 'A New Map of the Franco-Brazilian Border Dispute (1900)'. *Imago Mundi* 67 no. 2(2015): 229–242.

Ferretti, F., 'Arcangelo Ghisleri and the "Right to Barbarity": Geography and Anti-Colonialism in Italy in the Age of Empire (1875–1914)'. *Antipode* 48, no. 3(2016): 563–583. doi:10.1111/anti.12206.

Ferretti, F. and E. Castleton, 'Fédéralisme, identités nationales et critique des frontières naturelles: Pierre-Joseph Proudhon (1809–1865) géographe des "États-Unis d'Europe"', *Cybergeo, European Journal of Geography* (2016) [Retrieved from http://cybergeo.revues.org/27639] doi:10.4000/cybergeo.27639.

Ferretti, F. and P. Pelletier, 'En los orígenes de la geografía crítica. Espacialidades y relaciones de dominio en la obra de los geógrafos anarquistas Reclus, Kropotkin y Mechnikov'. *Germinal Revista de Estudios Libertarios* 11(2013): 57–72.

Fleming, M., 'Propaganda by the Deed. Terrorism and Anarchist Theory in Late Nineteenth-Century Europe'. In *Terrorism in Europe*, edited by A. Yonah and K.A. Myers, 8–28. London: Groom Helm, 1982.

Fleming, M., *The Geography of Freedom. The Odyssey of Élisée Reclus*. Montréal and New York: Black Rose Books, 1988.

Flint, C. and P. Taylor, *Political Geography*. Harlow: Pearson Education, 2007.

Flipo, F., 'Voyage dans la galaxie décroissante'. *Mouvements* 50(2007): 143–151.

Fones-Wolf, K. and R.L. Lewis, *Transnational West Virginia: Ethnic Communities and Economic Change, 1840 1940*. Morgantown: West Virginia University Press, 2011.

Foster, R.F., *Vivid Faces. The Revolutionary Generation in Ireland 1890–1923*. London: Allen Lane, 2014.

Foucault, M., 'Questions on Geography'. In *Space, Knowledge and Power: Foucault and Geography*, edited by J.W. Crampton and S. Elden, 172–182. Aldershot: Ashgate, 2007.

Foucault, M., *The Courage of Truth: The Government of Self and Others, Volume Two. Lectures at the Collège De France, 1983–1984*. Basingstoke: Palgrave Macmillan, 2011.

Fourier, C., *Des modifications à introduire dans l'architecture des villes, ouvres complètes* [Modifications to introduce in cities' architecture] Paris: Anthropos, 1967.

Funtowictz, S.O. and J.R. Ravetz, 'Science for the Post-Normal Age', *Futures* 25 (1993): 739–755.

Gallego, F.M., 'Deleuze y la filosofía de la ciencia'. *Revista Filosofía UIS* 9, no. 1 (2011): 61–80.

García, E., 'Degrowth, the Past, the Future, and the Human Nature'. *Futures* 44(2012): 546–552.

Gaspari, O. and P. Dogliani (eds), *L'Europa dei comuni: Origini e sviluppo del movimento communale europeo dall fine dell'Ottocento al secondo dopoguerra*. Rome: Donzelli, 2003.

Gellner, E., *Nations and Nationalism*. Ithaca, NY: Cornell University Press, 1983.

Genosko, G., *Félix Guattari: An Aberrant Introduction*. London: Bloomsbury Publishing, 2002.

Giblin, B., 'Reclus: un écologiste avant l'heure?' *Hérodote* 22(1981): 107–118.

Giblin, B. (ed.), *Élisée Reclus. El hombre y la Tierra*. Mexico City: Fondo de Cultura Económica, 1986.

Giblin, B., 'Reclus et les colonisations'. *Hérodote*117, no. 2(2005): 135–152.

Giullamón, A., *Ready for Revolution. The CNT Defense Committees in Barcelona 1933–38*. Oakland: AK Press, 2014.

Glassgold, P. (ed.), *Anarchy! An Anthology of Emma Goldman's Mother Earth*. Washington D.C.: Counterpoint, 2001.

Gledhill, J., 'Introduction: The Comparative Analysis of Social and Political Transitions'. In *State and Society. The Emergence and Development of Social Hierarchy and Political Centralization*, edited by J. Gledhill, B. Bender and M.T. Larsen, 1–30. New York: Routledge, 2005.

Gluck, M., *Popular Bohemia: Modernism and Urban Culture in Nineteenth-century Paris*. Cambridge, MA: Harvard University Press, 2005.

Gómez, J., Muñoz, J. and N. Cantero, *El pensamiento geográfico: estudio interpretativo y antología de textos: de Humboldt a las tendencias radicales* [Geographical thought: interpretative study and texts' anthology from Humboldt until radical tendencies]. Madrid: Alianza Editorial, 2002.

González-Ruibal, A., 'Postcolonialism and European Archaeology'. In *Handbook of Postcolonial Archaeology*, edited by J. Lydon and U.Z. Rizvi, 39–50. Walnut Creek: Left Coast, 2010.

Gordon, R.B., 'From Charcot to Charlot: Unconscious Imitation and Spectatorship in French Cabaret and Early Cinema'. *Critical Inquiry* 27, no. 3(2001): 515–549.

Gorostiza, S., March, H. and D. Sauri, 'Servicing Customers in Revolutionary Times: The Experience of the Collectivized Barcelona Water Company during the Spanish Civil War'. *Antipode* 45, no. 4(2012): 908–925. doi:10.1111/j.1467–8330.2012.01013.x.

Goudeau, E., *Dix ans de bohème*. Paris: Henry du Parc, 1888.

Goyens, T., *Beer and Revolution: the German Anarchist Movement in New York City, 1880–1914*. Urbana: University of Illinois Press, 2007.

Goyens, T., 'Social Space and the Practice of Anarchist History', *Rethinking History* 13, no. 4(2009): 439–457.

Goyens, T. (ed.), *Radical Gotham: Anarchism in New York City from Schwab's Saloon to Occupy Wall Street*. Urbana: University of Illinois Press, 2017.

Graham, R., *We Do not Fear Anarchy, We Invoke It! The First International and the Origins of the Anarchist Movement*, London: AK Press, 2015.

Gramsci, A., *The Modern Prince and Other Writings*, London: Lawrence and Wishart, 1957.

Grand-Carteret, J., *Raphael et Gambrinus, ou L'art dans la brasserie*. Paris, 1886.

Guillamón, A., *Ready for Revolution. The CNT Defense Committees in Barcelona 1933–1938*. Edinburgh: Edinburgh University Press, 2014.

Guillaume, J., 'Variétés. Bibliographie. *Nouvelle Géographie Universelle. La Terre et les Hommes*'. *Bulletin de la Fédération Jurassienne* 4, no. 24(1875): 3–4.

Häkli, J., 'In the Territory of Knowledge: State-centred Discourses and the Construction of Society' *Progress in Human Geography* 25, no. 3(2001): 403–422. doi:10.1191/030913201680191745.

Hall, P. and C. Ward, *Sociable Cities: The Legacy of Ebenezer Howard*. Chichester: John Wiley, 1998.

Halperin, J., *Félix Fénéon: Aesthete and Anarchist in fin-de-siècle Paris*. New Haven: Yale University Press, 1988.

Hardy, D. and C. Ward, *Arcadia for All: The Legacy of a Makeshift Landscape*. London: Mansell, 1984.

Harrison, R., 'Sydney and Beatrice Webb'. In *Socialism and the Intelligentsia*, edited by C. Levy, 35–89. London: Routledge, 1987/2017.

Harvey, D., 'Monument and Myth'. *Annals of the Association of American Geographers* 69, no. 3(1979): 362–381.

Harvey, D., 'On the History and Present Condition of Geography. An Historical Materialist Manifesto'. *The Professional Geographer* 36, no. 1(1984): 1–11. doi:10.1111/j.0033–0124.1984.00001.x.

Harvey, D., 'Cosmopolitanism and the Banality of Geographical Evils'. *Public Culture* 12, no. 2(2000): 529–564. doi:10.1215/08992363-12-2-529.

Harvey, D., '"Listen Anarchist!" A Personal Response to Simon Springer's "Why a Radical Geography must be Anarchist"', 10 June (2015) [Retrieved from http://davidharvey.org/2015/06/listen-anarchist-by-david-harvey/].

Hayat, S., 'Le Fédéralisme Proudhonien à l'Épreuve des Nationalités' [Proudhonian federalism and the challenge of nationalities]. In *Le Fédéralisme: le Retour?* edited by J. Cagiao y Conde, 41–58. Paris: Publications de la Société Proudhon. 2010.

Hechter, M., *Internal Colonialism: The Celtic Fringe in British National Development, 1536–1966*. London: Routledge, 1975.

Heckert, J., 'Sexuality as State Form'. In *Post-Anarchism: A Reader*, edited by D. Rousselle and S. Evren, 194–207. Ann Arbor/London: Pluto Press, 2011.

Hill, L. 'Human Geography and Archaeology: Strange Bedfellows?' *Progress in Human Geography* 39, no. 4(2015): 412–431. doi:10.1177/0309132514521482.

Hirsch, S. and L. van der Walt (eds), *Anarchism and Syndicalism in the Colonial and Postcolonial World, 1870–1940*. Leiden/Boston: Brill, 2010.

Hirsch, S., 'Peruvian Anarcho-Syndicalism: Adapting Transnational Influences and Forging Counterhegemonic Practices, 1905–1930'. In *Anarchism and Syndicalism in the Colonial and Postcolonial World, 1870–1940. The Praxis of National Liberation, Internationalism, and Social Revolution*, edited by S. Hirsch and L. van der Walt, 227–272. Leiden: Brill, 2014.

Hobsbawm, E., *The Invention of Tradition*. Cambridge: Cambridge University Press, 1983.

Hobsbawm, E., *The Age of Empire*. New York: Pantheon Books, 1987.

Hooson, D. (ed.), *Geography and National Identity*. Oxford: Blackwell, 1994.

Horowitz, M.C. (ed.), *New Dictionary of the History of Ideas*. New York: Scribners, 2005.

Hou, J., *Insurgent Public Space: Guerrilla Urbanism and the Remaking of Contemporary Cities*. London/New York: Routledge, 2010.

Hwang, D., 'Korean Anarchism before 1945: A Regional and Transnational Approach'. In *Anarchism and Syndicalism in the Colonial and Postcolonial World, 1870–1940. The Praxis of National Liberation, Internationalism, and Social Revolution* edited by S. Hirsch and L. van der Walt, 95–130. Leiden: Brill, 2014.

Hyman, E.W., 'Theatrical Terror: Attentats and Symbolist Spectacle'. *The Comparatist* 29(2005): 101–122.

Ince, A., 'In the Shell of the Old: Anarchist Geographies of Territorialisation'. *Antipode* 44, no. 5(2012): 1645–1666. doi:10.1111/j.1467–8330.2012.01029.x.

Ince, A., 'Black Flag Mapping. Emerging Themes in Anarchist Geography'. In *The Anarchist Imagination: Anarchism Encounters the Humanities and Social Sciences*, edited by C. Levy and S. Newman. London, 2017.

Ince, A. and G. Barrera de la Torre, 'For Post-Statist Geographies'. *Political Geography* 55, no. 1(2016). 10–19.

Ishill, J. (ed.), *Elisée and Elie Reclus: In Memoriam*. Berkeley Heights: Oriole Press, 1927.

Jessop, B., *State Power*. Cambridge: Polity, 2007.

Joll, J., *The Anarchists*. London: Eyre & Spottiswoode, 1964.

Jonas, R., 'Sacred Tourism and Secular Pilgrimage: Montmartre and the Basilica of Sacré Coeur'. In *Montmartre and the Making of Mass Culture*, edited by G. Weisberg, 94–119. New Brunswick, NJ: Rutgers University Press, 2001.

Jones, L., *Sad Clowns and Pale Pierrots: Literature and the Popular Comic Arts in 19th-Century France*. Lexington, KY: French Forum, 1984.

Jones, R., 'State Encounters'. *Environment and Planning D: Society and Space* 30, no. 5(2012): 805–821. doi:10.1068/d9110.

Jones, R., Pykett, J. and M. Whitehead, 'Governing Temptation: Changing Behaviour in an Age of Libertarian Paternalism'. *Progress in Human Geography* 35, no. 4(2010): 483–501. http://dx.doi.org/10.1177/0309132510385741.

Jones, T., *More Powerful than Dynamite. Radicals, Plutocrats, Progressives, and New York's Year of Anarchy*. New York: Walker & Company, 2012.

Jud, P., *Elisée Reclus und Charles Perron, Schöpfer der "Nouvelle Geographie Universelle"* [Elisée Reclus and Charles Perron, authors of the Nouvelle Géographie Universelle]. Zurich: Verlag, 1987.

Jun, N., *Anarchism and Political Modernity*. London: Continuum, 2012.

Kallis, G. and H. March, 'Imaginaries of Hope: The Utopianism of Degrowth'. *Annals of the Association of American Geographers*, 105(2014): 360–368. doi:10.1080/00045608.2014.973803.

Kallis, G., Demaria, F. and G. D'Alisa, 'Introduction: Degrowth'. In *Degrowth: A Vocabulary for a New Era*, edited by G. D'Alisa, F. Demaria, and G. Kallis, 1–17. New York: Routledge, 2015.

Kanngieser, A., *Experimental Politics and the Making of Worlds*. Farnham: Ashgate, 2013.

Kaplan, T., *Red City. Blue Period. Social Movement in Picasso's Barcelona*. Berkeley: University of California Press, 1992.

Kearns, G., 'The Political Pivot of Geography'. *The Geographical Journal* 170(2004): 337–346. doi:10.1111/j.0016–7398.2004.00135.x.

Kearns, G., *Geopolitics and Empire. The Legacy of Halford Mackinder*. Oxford: Oxford University Press, 2009.

Khuri-Makdisi, I., *The Eastern Mediterranean and the Making of the Global Radicalism, 1860–1914*. Berkeley: University of California Press, 2010.

Kinna, R., *Kropotkin: Reviewing the Classical Anarchist Tradition*. Edinburgh: Edinburgh University Press, 2016.

Kinna, R., 'Kropotkin's Theory of the State: A Transnational Approach'. In *Reassessing the Transnational Turn. Scales of Analysis in Anarchist and Syndicalist Studies*, edited by C. Bantman and B. Altena, 43–61. London: Routledge, 2015.

Knapp, M., Flach, A. and E. Ayboga, *Revolution in Rojava: Democratic Autonomy and Women's Liberation in Syrian Kurdistan*. London: Pluto Press, 2016.

Konishi, S., *Anarchist Modernity: Cooperatism and Japanese–Russian Intellectual Relations in Modern Japan*. Cambridge: Harvard University Press, 2013.

Koselleck, R. 'Einleitung' [Introduction]. In *Geschichtliche Grundbegriffe. Historisches Lexikon zur politisch-sozialen Sprache in Deutschland, vol. 1* [Historical basic concepts. Historical lexicon of politico-social language in Germany], edited by R. Koselleck, W. Conze and O. Brunner, XIII–XXVII. Stuttgart: Cotta, 1972.

Krenak, A., 'O eterno retorno do encontro'. 1999 [Retrieved from http://www.geledes.org.br/narrativa-krenak-o-eterno-retorno-do-encontro/ on 25 August 2016].

Kropotkin, P., 'Finland: A Rising Nationality'. *The Nineteenth Century* 17(1885): 527–546.

Kropotkin, P., 'What Geography Ought to Be'. *The Nineteenth Century* 18(1885): 940–956.

Kropotkin, P., 'On the Teaching of Physiography'. *The Geographical Journal* 2, no. 4(1893) 350–359.

Kropotkin, P., *Fields, Factories and Workshops*. London: Ward, 1899.

Kropotkin, P., *Memoirs of a Revolutionist*. London: Smith, 1899.

Kropotkin, P., *Mutual Aid: A Factor of Evolution*. London: Heinemann, 1902.

Kropotkin, P., *Modern Science and Anarchism*. Philadelphia: The Social Science Club, 1903.

Kropotkin, P., 'What Geography Ought to Be'. *Antipode* 10(1978): 6–15.

Kuipers, G., *Good Humor, Bad Taste: A Sociology of the Joke*. Berlin & New York: Mouton de Gruyter, 2006.

Lacoste, Y., 'Elisée Reclus, une très large conception de la géographicité et une bienveillante géopolitique'. *Hérodote* 117, no. 2 (2005): 29–52.

Lacoste, Y., 'Hérodote et Reclus'. *Hérodote* 117, no. 2(2005), 5–8.

Lacoste, Y., *La géographie, ça sert d'abord à faire la guerre* [Geography serves first to make war] Paris: Maspero, 1976.

Landauer, G., *Revolution and Other Writings: a Political Reader*. London: Merlin Press, 2010.

Landauer, G., *Escepticismo y mística. Aproximaciones a la crítica del lenguaje de Mauthner*. [Scepticism and mystics: joining Mauthner's critique of language]. México: Herder, 2015.

Landauer, G., 'Weak Statesmen, Weaker People'. In *Revolution and Other Writings*, edited by G. Kuhn, 213–214. Oakland, CA: PM Press, 2010.

Latouche, S., *Survivre au développement* [Surviving development]. Paris: Mille et Une Nuits, 2004.

Latouche, S., *Le pari de la décroissance*. Paris: Fayard, 2006.

Latouche, S., *Farewell to Growth*. Cambridge-Malden: Polity Press, 2009.

Latouche, S., 'Can the Left Escape Economism'. *Capitalism, Nature, Socialism* 23, no. 1(2012): 74–78. doi:10.1080/10455752.2011.648841.

Latouche, S., 'Decolonization of Imaginary'. In *Degrowth: A Vocabulary for a New Era*, edited by G. D'Alisa, F. Demaria, and G. Kallis, 117–120. New York: Routledge, 2015.

Law, J., 'What's Wrong with a One-world World?'. 2011 [Retrieved from http://www. heterogeneities.net/publications/Law2011WhatsWrongWithAOneWorldWorld.pdf].

Lefebvre, H., *Pyrénées*. Lausanne: Éditions Rencontre, 1965.

Lefebvre, H., *La proclamation de la Commune* [The Commune's Proclamation]. Paris: Gallimard, 1965.

Lefebvre, H., *La production de l'espace* [The production of space]. Paris: Anthropos, 1974.

Lefebvre, H., *A revolução urbana* [The urban revolution]. Belo Horizonte: UFMG, 1999.

Lefebvre, H., *Espaço e política* [Space and politics]. Belo Horizonte: UFMG, 2007.

Lefebvre, H., *La somme et le reste* [The sum and the rest]. Paris: Anthropos, 2009.

Lefort, I. and P. Pelletier (eds), *Élisée Reclus et nos géographies. Textes et prétextes* [Élisée Reclus et our geographies. Textes et pretexts]. Paris: Noir et Rouge, 2013.

Lehardy, J., 'Montmartre'. *Le Chat Noir*, 14 January (1882).

Lehning, A., 'Michel Bakounine et le Risorgimento tradito'. *Bollettino del Museo del Risorgimento di Bologna* 17/19 (1972–1974): 266–292.

Leighton, P., *The Liberation of Painting. Modernism and Anarchism in Avant-Guerre Paris*. Chicago: University of Chicago Press, 2013.

Lepetit, B., *Por uma nova história urbana* [For a new urban history]. São Paulo: Edusp, 2003.

Levitt, P. and N. Glick Schiller, 'Conceptualizing Simultaneity: A Transnational Social Field Perspective on Society'. *The International Migration Review* 38, no. 3 (2004): 1002–1039. doi:10.1111/j.1747–7379.2004.tb00227.x.

Levy, C., 'The Rooted Cosmopolitan: Errico Malatesta, Syndicalism, Transnationalism and the International Labour Movement'. In *New Perspectives on Anarchism, Labour & Syndicalism: The Individual, the National and the Transnational*, edited by D. Berry and C. Bantman, Newcastle upon Tyne: Cambridge Scholars Publishing, 2010.

Levy, C., 'Anarchism, Internationalism and Nationalism in Europe, 1860–1939'. *Australian Journal of Politics and History* 50, no. 3(2004): 330–342. doi:10.1111/j.1467–8497.2004.00337.x.

Levy, C., 'Italian Anarchism, 1870–1926'. In *For Anarchism. History, Theory, and Practice*, edited by D. Goodway, 25–78. London: Routledge, 1989.

Levy, C., 'Charisma and Social Movements: Errico Malatesta and Italian Anarchism'. *Modern Italy* 3, no. 2(1998): 205–217.

Levy, C., 'Max Weber, Anarchism and Libertarian Culture: Personality and Power Politics'. In *Max Weber and the Culture of Anarchy*, edited by S. Whimster, 83–109. Basingstoke: Macmillan, 1999.

Levy, C., *Gramsci and the Anarchists*. Oxford: Berg, 1999.

Levy, C., 'Social Histories of Anarchism'. *Journal for the Study of Radicalism* 4, no. 2 (2010): 1–44. doi:10.1353/jsr.2010.0003.

Levy, C., 'Anarchism and Cosmopolitanism'. *Journal of Political Ideologies* 16, no. 3 (2011): 265–278. doi:10.1080/13569317.2011.607293.

Levy, C. (ed.), *Colin Ward. Life, Times and Thought*. London: Lawrence and Wishart, 2013.

Levy, C., 'New Anarchism and the City from the Cold War to the Occupy and Square Movements'. Ms, forthcoming.

Linebaugh, P. and Redikey, M., *The Many Headed Hydra: Sailors, Slaves and Commoners and the Hidden History of the Revolutionary Atlantic*. London: Verso, 2000.

Lisbonne, M., 'La Taverne du Bagne'. *Gazette du Bagne*1, 15 November (1885).

Livingstone, D.N., *The Geographical Tradition*. Oxford and Cambridge: Wiley, 1992.

Livingstone, D.N., 'Science, Text and Space: Thoughts on the Geography of Reading'. *Transactions of the Institute of British Geographers* 30, no. 4(2005): 391–401. doi:10.1111/j.1475–5661.2005.00179.x.

Livingstone, D.N. and C.W.J. Withers (eds), *Geography and Revolution*. Chicago: The University of Chicago Press, 2005.

Livingstone, D.N. and C.W.J. Withers (eds), *Geographies of Nineteenth-Century Science*. Chicago: The University of Chicago Press, 2011.

Llorente, J.M., 'Colonialismo y geografía en España en el último cuarto del siglo XIX. Auge y descrédito de la geografía colonial' [Colonialism and geography in the last quarter of the 19th century: the rise and discrediting of colonial geography]. *Eria* 15 (1988): 51–76.

Lloyd, N., *Forgotten Places: Barcelona and the Spanish Civil War*. London: Create Space, 2015.

Lockyer, S. and M. Pickering (eds), *Beyond a Joke: The Limits of Humour*. Basingstoke: Palgrave Macmillan, 2005.

Lopes, M., 'Élisée Reclus e o Brasil'. *GEOgraphia* 11, no. 2(2009): 160–175 [Retrieved from http://www.uff.br/geographia/ojs/index.php/geographia/issue/view/23].

Lord, E.A.M., Trenor, J.J.D. and S.J. Barrows, *The Italian in America*. New York: B. F. Buck and Company, 1906.

Lovell, S., 'Nihilism, Russian'. In *Routledge Encyclopaedia of Philosophy*, edited by E. Craig. London: Routledge, 1998.

Lydon, J. and U.Z. Rizvi. *Handbook of Postcolonial Archaeology*. Walnut Creek: Left Coast, 2010.

Mackinder, H.J., 'Reclus' "Universal Geography"'. *The Geographical Journal* 4, no. 2(1894): 158–160.

Mackinder, H.J., *Britain and the British Seas*. London: D. Appleton & Company, 1902.

'Mackinder, Mr., Ravenstein, Mr., Herbertson, Dr., Kropotkin, P., Andrews, Mr., Sanderson, C. and E. Reclus. On Spherical Maps and Reliefs: Discussion'. *The Geographical Journal* 22, no. 3(1903): 294–299.

MacLaughlin, J., *Kropotkin and the Anarchist Intellectual Tradition*. London: Pluto Press, 2016.

Malatesta, E., 'La Comune di Parigi e gli anarchici' [The Paris Commune and anarchy]. *La Settimana Sanguinosa*, 18–24 March (Special Issue) (1903).

Malatesta, E., 'La propaganda del compagno E. Malatesta' [The propaganda of comrade E. Malatesta]. *Umanità Nova*, 16 September (1920).

Malatesta, E., 'Sur Pierre Kropotkine. Souvenirs et critiques d'un de ses vieux amis' [On Pierre Kropotkin. Memories and critiques of one of his old friends]. *StudiSociali* 3(1931): 368–379.

Malatesta, E., *'Verso l'anarchia': Malatesta in America, 1899–1900* ['Towards anarchy': Malatesta in America, 1899–1900], edited by D. Turcato. Milan: Zero in Condotta and Ragusa-La Fiaccola, 2012.

Malatesta, E., *'Lo sciopero armato': Il lungo esilio londinese, 1900–1913* ['The armed strike': The long exile in London 1900–1913], ed. D. Turcato. Milan: Zero in Condotta and Ragusa-La Fiaccola, 2015.

Manfredonia, G. (ed.), *Les anarchistes et la Révolution française* [Anarchists and the French Revolution]. Paris: Éditions du Monde Libertaire, 1990.

Mantovani, V., *Mazurka blu: la strage del Diana* [Mazurka blu. The Diana's slaughter]. Milan: Rusconi, 1979.

Marcus, J. and G.M. Feinman, 'Introduction'. In *Archaic States*, edited by J. Marcus and G.M. Feinman, 3–17. Santa Fe: School of America Research Press, 1998.

Marcus, J. and G.M. Feinman, *Archaic States*. Santa Fe: School of America Research Press, 1998.

Marshall, P., *Demanding the Impossible: a History of Anarchism*. Oakland: PM Press, 2010.

Marston, S., 'Space, Culture, State: Uneven Developments in Political Geography'. *Political Geography* 23, no. 1(2004): 1–16. doi:10.1016/j.polgeo.2003.09.006.

Marston, S., JonesIII, J.P. and K. Woodward. 'Human Geography without Scale'. *Transactions of the Institute of British Geographers* 30, no. 4(2005): 416–432. doi:10.1111/j.1475–5661.2005.00180.x.

Martin, L.L. and M.L. Mitchelson, 'Geographies of Detention and Imprisonment: Interrogating Spatial Practices of Confinement, Discipline, Law and State Power'. *Geography Compass* 3, no. 1(2009): 459–477. doi:10.1111/j.1749-8198.2008.00196.x.

Martínez Luna, J., 'Comunalizar la vida toda', 2003 [Retrieved from http://www.nasaa cin.org/attachments/article/6415/comunalicemos%20la%20vida%20toda.pdf on 4 August 2016].

Martínez Luna, J., *Eso que llaman comunalidad* [*What is called communality*]. México: Culturas Populares, CONACULTA/Secretaría de Cultura, Gobierno de Oaxaca/ Fundación Alfredo Harp Helú Oaxaca, 2009.

Martínez-Alier, J., *The Environmentalism of the Poor*. Massachusetts: Edward Elgar, 2002.

Martínez-Alier, J., 'Neo-Malthusians'. In *Degrowth: A Vocabulary for a New Era*, edited by G. D'Alisa, F. Demaria, and G. Kallis, 125–128. New York: Routledge, 2015.

Martínez-Alier, J., Pascual, U., Vivien, F. and E. Zaccai. 'Sustainable De-growth: Mapping the Context, Criticisms and Future Prospects of an Emergent Paradigm'. *Ecological Economics* 69, no. 9(2010): 1741–1747. doi:10.1016/j.ecolecon.2010.04.017.

Masini, P.C., *Eresie dell'Ottocento: Alle sorgenti laiche, umaniste e libertarie della democrazia italiana* [Nineteenth century heresies: lay, humanist and libertarian sources of Italian democracy]. Milan: Editoriale nuova, 1978.

Masjuan, E., *La ecología humana en el anarquismo ibérico: urbanismo orgánico o ecológico, neomalthusianismo y naturismo social* [Human ecology in Iberian anarchism. Organic or ecological urbanism, neomalthusianism and social naturism]. Barcelona: Icaria Editorial, 2000.

Massey, D., *For Space*. London: Sage, 2005.

Mavor, J., *My Windows on the Street of the World*. London-Toronto: J. M. Dent & Sons Limited, 1923.

Maxwell, B. and R. Craib (eds), *No Gods No Master No Peripheries. Global Anarchisms*. Oakland: PM Press, 2015.

McAnany, P.A. and N. Yoffee, 'Why we Question Collapse and Study Human Resilience, Ecological Vulnerability, and the Aftermath of Empire'. In *Questioning Collapse: Human Resilience, Ecological Vulnerability, and the Aftermath of Empire*, edited by P. A. McAnany and N. Yoffee, 1–17. Cambridge: Cambridge University Press, 2010.

McAnany, P.A. and N. Yoffee (eds), *Questioning Collapse: Human Resilience, Ecological Vulnerability, and the Aftermath of Empire*. Cambridge: Cambridge University Press, 2010.

McGuire, R.H., *Archaeology as Political Action*. Berkeley and Los Angeles: University of California Press, 2008.

Merriman, J., *Massacre. The Life and Death of the Paris Commune*. New York: Basic Books, 2014.

Mitchell, D., *Cultural Geography*. Oxford: Blackwell, 2000.

Miessen, M., *The Nightmare of Participation. Cross Benching Praxis as a Mode of Criticality*. Berlin: Sternberg Press, 2010.

Miessen, M. and S. Basar, *Did Someone Say Participate: An Atlas of Spatial Practice*. Cambridge, MA: MIT Press, 2006.

Miller, M.A., *Kropotkin*. Chicago: Chicago University Press, 1976.

Miyahiro, M.A., *O Brasil de Élisée Reclus: Territorio e sociedade em fins do século XIX* [The Brazil of Élisée Reclus: territory and society in the late nineteenth century]. São Paulo: São Paulo University, Unpublished Master of Arts dissertation, 2011.

Moisio, S. and A. Paasi, 'Beyond State-centricity: Geopolitics of Changing State Spaces'. *Geopolitics* 18, no. 2(2013): 255–266. doi:10.1080/14650045.2012.738729.

Monte Mor, R., 'Extended Urbanization and Settlement Patterns in Brazil: An Environmental Approach'. In: *Implosions/Explosions: Toward a Study of Planetary Urbanization*, edited by N. Brenner, 109–120. Berlin: Jovis Verlag GmbH, 2014.

Moréas, J., 'Le Symbolisme'. *Le Figaro, Supplément littéraire*, 18 September (1886): 1–2.

Morehart, C.T., 'What if the Aztec Empire Never Existed? The Prerequisites of Empire and the Politics of Plausible Alternative Histories'. *American Anthropologist* 114, no. 2(2012): 267–281. doi:10.1111/j.1548-1433.2012.01424.x.

Mormino, G.R. and G.E. Pozzetta, *The Immigrant World of Ybor City Italians and Their Latin Neighbors in Tampa, 1885–1985*. Gainesville: University Press of Florida, 1998.

Mormino, G.R. and G.E. Pozzetta, 'The Radical World of Ybor City, Florida'. In *The Lost World of Italian American Radicalism: Politics, Labor, and Culture*, edited by P.V. Cannistraro and G. Meyer, 253–263. Westport, Connecticut: Praeger, 2003.

Morris, B., *The Anarchist Geographer. An Introduction to the Life of Peter Kropotkin*. Somerset: The Genge Press, 2012.

Mould, O., *Urban Subversion and the Creative City*. London: Routledge, 2015.

Mountz, A., 'Human Smuggling, the Transnational Imaginary, and Everyday Geographies of the Nation-state'. *Antipode* 35, no. 3(2003): 622–644. doi:10.1111/1467-8330.00342.

Mountz, A., 'Political Geography I: Reconfiguring Geographies of Sovereignty'. *Progress in Human Geography* 37, no. 6(2013): 829–841. doi:10.1177/0309132513479076.

Mumford, L., *The City in History*. New York: Harcourt, Brace and World, 1961.

Mundy, J. (ed.), *Duchamp Man Ray Picabia*. London: Tate Publishing, 2008.

Mysyrowicz, L., 'Agents Secrets Tsaristes et Révolutionnaires Russes à Genève 1879–1903'. *Schweizerische Zeitschrift für Geschichte/Revue Suisse d'histoire/Rivista storica svizzera* 23(1973): 29–48.

Naylor, S., 'Historical Geography: Knowledge, in Place and on the Move'. *Progress in Human Geography* 29, no. 5(2005): 626–634. doi:10.1191/0309132505ph573pr.

Naumann, F. A., 'Aesthetic Anarchism', in Mundy J. (ed.), *Duchamp Man Ray Picabia*. London: Tate Publishing, 2008.

Negri, A., *O Poder Constituinte. Ensaio sobre as alternativas da modernidade* [The constituent power. Essay on the alternatives of modernity]. Rio de Janeiro: DP&A, 2002.

Negri, A., 'Dispositivo metrópole. A multidão e a metrópole'. *Revista Lugar Comum. Estudos de Mídia, Cultura e Democracia* 25–26(2008): 201–208.

Negri, A. and M. Hardt, *Multidão. Guerra e democracia na era do império* [Crowd. War and democracy in the age of empire]. Rio de Janeiro: Record, 2005.

Nettlau, M., *Geschichte der Anarchie* [History of anarchy]. Berlin: Der Syndikalist, 1925.

Nettlau, M., *Elisée Reclus Anarchist und Gelehrter (1830–1905)*. Berlin: Der Syndikalist, 1928.

Nettlau, M., *Eliseo Reclus: vida de un sabio justo y rebelde, Vol. II* [Élisée Reclus: life of a just and rebellious sage]. Barcelona: Ediciones de la Revista Blanca, 1930.

Nettlau, M., *La Première Internationale en Espagne (1868–1888)* [The First International in Spain (1868–1888)]. Dordrecht: Redider Publishing Company, 1969.

Nietzsche, F., *Thus Spake Zarathustra*. Mineola, NY: Dover Publications, 1999.

Nisbet, R., *History of the Idea of Progress*. New York: Basic Books, 1980.

Ó Donghaile, D., *Blasted Literature. Victorian Political Fiction and the Shock of Modernism*. Edinburgh: University of Edinburgh Press, 2011.

Olwig, K.R., 'Historical Geography and the Society/Nature "Problematic": The Perspective of J.F. Schow, G.P. Marsh and E. Reclus'. *Journal of Historical Geography* 6, no. 1(1980): 29–45.

Orwell, G., *Homage to Catalonia*. Harmondsworth: Penguin Books, 1966.

Osterhammel, J., 'Die Wiederkehr des Raumes. Geopolitik, Geohistorie und historische Geographie'. *Neue Politische Literatur*, 43(1998): 374–397.

Painter, J., 'Prosaic Geographies of the State'. *Political Geography* 25, no. 7(2006): 752–774. doi:10.1016/j.polgeo.2006. 07. 004.

Paperno, I., *Chernyshevsky and the Age of Realism. A Study in the Semiotics of Behavior*. Standford: Stanford University Press, 1988.

Pelletier, P., *L'imposture écologiste* [The environmentalist imposture]. Reclus: Montpellier: 1993.

Pelletier, P., *Géographie et anarchie: Reclus, Kropotkine, Metchnikoff* [Geography and anarchy: Reclus, Kropotkin, Metchnikoff]. Paris: Éditions du Monde Libertaire, 2013.

Pelletier, P., *Albert Camus, Élisée Reclus et l'Algérie. Les 'indigènes de l'univers'* [Albert Camus, Élisée Reclus and Algeria. The natives of the universe]. Paris: Le Cavalier Bleu Editions, 2015.

Pelletier, P., 'Élisée Reclus et la mésologie'. *Colloque Retour des territoires, renouveau de la mésologie*, Università di Corsica, Corte, 26–27 March, 2015.

Pernicone, N., *Carlo Tresca: Portrait of a Rebel*. Edinburgh: AK Press, 2010.

Pettit, P., *Republicanism: A Theory of Freedom and Government*. Oxford: Clarendon Press, 1997.

Pezzarossi, G., 'A Spectral Haunting of Society: Longue Durée Archaeologies of Capitalism and Antimarkets in Colonial Guatemala'. In *Historical Archaeologies of Capitalism*, edited by M.P. Leone and J.E. Knauf, 345–373. New York: Springer, 2015.

Philo, C., 'Neglected Rural Geographies: A Review'. *Journal of Rural Studies* 8, no. 2 (1992): 193–207. doi:10.1016/0743-0167(92)90077-J.

Picon, A., 'Racionalidade técnica e utopia: a gênese da haussmanização'. In *Cidades capitais do século XIX: racionalidade, cosmopolitismo e transferência de modelos* [Capital cities of the nineteenth century: rationality, cosmopolitanism and transference of models], edited by H.A. Salgueiro, 65–102. São Paulo: EDUSP, 2001.

Pirumova, N.M., *Pëtr Alekseevich Kropotkin*, Moscow: Nayka, 1972.

Poggi, C., *Inventing Futurism. The Art and Politics of Artificial Optimism*. Princeton: Princeton University Press, 2009.

Powell, C. and G. Paton (eds), *Humour in Society: Resistance and Control*. Basingstoke: Macmillan, 1988.

Powell, W.E., 'Former Mining Communities of the Cherokee-Crawford Coal Field of Southeastern Kansas'. *Kansas Historical Quarterly* 38, no. 2(1972): 187–199.

Préposiet, J., *Histoire de l'anarchisme* [History of anarchism]. Paris: Tallandier, 2005.

Preston, P., *The Spanish Holocaust. Inquisition and Extermination in Twentieth-Century Spain*. London: Harper Press, 2012.

Prudovsky, G., 'Can We Ascribe to Past Thinkers Concepts They Had No Linguistic Means to Express?' *History and Theory* 36, no. 1(1997): 15–31. doi:10.1111/0018–2656.00002.

Purchase, G., *Anarchism and Environmental Survival*. Tucson, Arizona: See Sharp Press, 1994.

Quattrochi-Woisson, D., *Juan Bautista Alberdi y la independencia Argentina. La fuerza del pensamiento y de la escritura* [Juan Bautista Alberdi and the Argentinian Independence: the power of thought and writing]. Buenos Aires: Universidad Nacional de Quilmes, 2012.

Ramirez Palacios, D., *Élisée Reclus e a Geografia da Colômbia: Cartografia de uma interseção* [Élisée Reclus and the geography of Colombia: mapping an intersection]. São Paulo: São Paulo University, Unpublished Master of Arts dissertation, 2010.

Ramnath, M., *Decolonizing Anarchism: An Antiauthoritarian History of India's Liberation Struggle*. London: AK Press, 2011.

Rapp, J.A., *Daoism and Anarchism: Critiques of State Autonomy in Ancient and Modern China*. London: Continuum, 2012.

Reclus, E., 'Le Brésil et la colonisation, I. Le bassin des Amazones et les Indiens'. *La Revue des Deux Mondes* 39(1862): 375–414.

Reclus, E., 'Le Brésil et la colonisation, II. Les provinces du littoral, les noirs et les colonies allemandes'. *La Revue des Deux Mondes* 40(1862): 930–959.

Reclus, E., 'L'homme et la nature: de l'action humaine sur la géographie physique [recensione a: G.P. Marsh, Man and nature]'. *La Revue des Deux Mondes* 54(1864): 762–771.

Reclus, E., 'La guerre de l'Uruguay et les républiques de La Plata'. *La Revue des Deux Mondes* 55(1865):967–997.

Reclus, E., 'Du sentiment de la nature dans les sociétés modernes'. *La Revue des Deux Mondes* 63(1866): 352–381.

Reclus, E., 'L'élection présidentielle de la Plata et la guerre du Paraguay'. *La Revue des Deux Mondes* 76(1868):891–910.

Reclus, E., *La Terre: description des phénomènes de la vie du globe* [The Earth: description of the phenomena of the life of the globe], 2 vols. Paris: Hachette, 1868–1869.

Reclus, E., *Histoire d'un ruisseau* [The history of a creek]. Paris: Hachette, 1869.

Reclus, E., *The Ocean, Atmosphere and Life*. New York: Harper and Brothers, 1873.

Reclus, E., *Nouvelle Géographie Universelle. La terre et les hommes* [Trans. as *The Earth and Its Inhabitants: The Universal Geography*], 19 vols. Paris: Hachette, 1876–1894.

Reclus, E., *Histoire d'une montagne* [The history of a mountain]. Paris: Hetzel, 1880.

Reclus, E., *The Earth and Its Inhabitants: The Universal Geography*, 19 vols. New York: Appleton and Co., 1876–1894.

Reclus, E., 'A mon frère le paysan'. *L'Idée Libre* 20(1893): 2–16.

Reclus, E., 'L'anarchie'. *Les Temps nouveaux* 2(1895): 7–23.

Reclus, E., 'La grande famille', *Le Magazine International*, January (1896): 8–12.

Reclus, E., 'The progress of mankind'. *The Contemporary Review* 70(1896): 761–783.

Reclus, E., *L'évolution, la révolution et l'idéal anarchique* [Evolution, revolution and the anarchist ideal]. Paris: Stock, 1897.

Reclus, E., 'Léopold de Saussurre, *Psychologie de la colonisation française dans ses rapports avec les sociétés indigènes*, Paris, Alcan, 1899'. *L'Humanité Nouvelle* 26 (1899):246–248.

Reclus, E., *Estados Unidos do Brasil, Geographia, Ethnographia, Estatistica, por Élisée Reclus. Tradução e Breves Notas de B. F. Ramiz Galvão, e Annotações Sobre o Território Contestado pelo Barão do Rio Branco* [United States of Brazil, ethnographic geography and statistics by Élisée Reclus. Translation and notes by B. F. Ramiz Galvão, and annotations on the contested territory of the Baron of Rio Branco]. Rio de Janeiro: Garnier, 1900.

Reclus, E., 'L'enseignement de la géographie'. *Bulletin de la Société Belge d'Astronomie* 11(1903): 5–11.

Reclus, E., *L'Homme et la Terre* [Man and the Earth], 6 vols. Paris: Librairie Universelle, 1905–1908.

Reclus, E., *Correspondencia 1850–1905* [*Correspondence 1850–1905*]. Buenos Aires: Ediciones Imán, 1943.

Reclus, P., *Les frères Élie et Élisée Reclus. Ou du protestantisme à l'anarchisme* [The Brothers Reclus. Or from Protestantism to anarchism]. Paris: Les Amis d'Élisée Reclus, 1964.

Ridanpää, J., 'Geopolitics of Humour: The Muhammed Cartoon Crisis and the Kaltio Comic Strip Episode in Finland'. *Geopolitics* 14, no. 4(2009): 729–749. doi:10.1080/14650040903141372.

Rider, N., 'The Practice of Direct Action: the Barcelona Rent Strike of 1931'. In *For Anarchism. History, Theory, and Practice*, edited by D. Goodway, 79–108. London: Routledge, 1989.

Ridoux, N., *La décroissance pour tous* [Degrowth for all]. Lyon: Parangon/Vs, 2006.

Riechmann, J., *Biomímesis: ensayos sobre imitación de la naturaleza, ecosocialismo y autocontención* [*Biomimicry: essays on imitation of nature, ecosocialism and self-restraint*]. Madrid: Los Libros de la Catarata, 2006.

Rist, G., *The History of Development: From Western Origins to Global Faith*. London: Zed Books, 2003.

Ritter, J., Gründer, K. and G. Gottfried (eds), *Historisches Wörterbuch der Philosophie* [*Historical dictionary of philosophy*]. Darmstadt: Schwabe, 1971–2007.

Robic, M.C. (ed.), *Le Tableau de la Géographie de la France de Paul Vidal de la Blache: dans le Labyrinthe des Formes* [The Tableau of Geography of France of Paul Vidal de la Blache: the labyrinth of forms]. Paris: CTHS, 2000.

Robinson, E.H., 'The Distinction between State and Government'. *Geography Compass* 7, no. 8(2013): 556–566. doi:10.1111/gec3.12065.

Rogers, J.D., 'The Contingencies of State Formation in Eastern Inner Asia'. *Asian Perspectives* 46, no. 2(2007): 249–274. doi:10.1353/asi.2007.0017.

Romani, C., 'Da Biblioteca Popular à Escola Moderna. Breve história da ciência e educação libertária na América do Sul'. *Educacão Libertaria* 1(2006): 87–100.

Romani, C., *Aquí Começa o Brasil! Histórias das Gentes e dos Poderes na Fronteira do Oiapoque* [Here begins Brazil! Stories of people and the powers on the border of Oiapoque]. Rio de Janeiro: Editora Multifoco, 2013.

Rosenthal, A., 'Moving between the Global and the Local: The Industrial Workers of the World and their Press in Latin America'. In *Defiance of Boundaries. Anarchism in Latin American History*, edited by G. de Laforcade and K. Shaffer, 72–93. Gainesville: University Press of Florida, 2015.

Ross, K., *May 68 and its Afterlives*. Chicago: University of Chicago Press, 2002.

Ross, K., *The Emergence of Social Space. Rimbaud and the Paris Commune*. London and New York: Routledge, 2008.

Ross, K., *Communal Luxury. The Political Imaginary of the Paris Commune*. London and New York: Verso, 2015.

Rothen, E., 'Elisée Reclus' Optimism'. In *Elisée and Elie Reclus: In Memoriam*, edited by J. Ishill. Berkeley Heights, N.J.: Oriole Press, 1927.

Routledge, P., 'Sensuous Solidarities: Emotion, Politics and Performance in the Clandestine Insurgent Rebel Clown Army'. *Antipode* 44, no. 2(2012): 428–452. doi:10.1111/j.1467-8330.2010.00862.x.

Ryley, P., *Making Another World Possible. Anarchism, Anti-Capitalism and Ecology in Late 19th and early 20th Century Britain*. London: Bloomsbury, 2013.

Salaris, C., *Alla festa della rivoluzione. Artisti e literati con D'Annunzio a Fiume* [The party of the revolution. Artists and writers with D'Annunzio in Fiume]. Bologna: Il Mulino, 2008.

Salis, R., 'Poster for the Montmartre Municipal Election'. Paris, 1884. Cited in M. Wilson 'Portrait of the Artist as a Louis XIII Chair'. In *Montmartre and the Making of Mass Culture*, edited by G. Weisberg, 200. New Brunswick, NJ: Rutgers University Press, 2011.

Salgueiro, H.A., *Cidades capitais do século XIX: racionalidade, cosmopolitismo e transferência de modelos* [Capital cities of the nineteenth century: rationality, cosmopolitanism and transference of models]. São Paulo: EDUSP, 2001.

Sandos, J.A., *Rebellion in the Borderlands: Anarchism and the Plan of San Diego, 1860–1923*. Tulsa: University of Oklahoma Press, 1992.

Santos, B.S., *Refundación del Estado en América Latina. Perspectivas desde una epistemología del Sur* [Founding the State in Latin America. Perspectives from an epistemology of the South]. Lima: Instituto Internacional de Derecho y Sociedad y Programa Democracia y Transformación Global, 2012.

Sbert, J.M., 'Progress'. In *The Development Dictionary. A Guide to Knowledge as Power*, edited by W. Sachs, 212–227. London and New York: Zed Books, 1992.

Scanlan, J.P., 'Russian Materialism. "The 1860s"'. In *Concise Routledge Encyclopedia of Philosophy*, edited by E. Craig. London and New York: Routledge, 2000.

Schmidt, W., *Nihilismus und Nihilisten. Untersuchungen zur Typisierung im russischen Roman der zweiten Hälfte des neunzehnten Jahrhunderts* [Nihilism and nihilists. Investigations on typology in the Russian novel of the second half of the nineteenth century]. Munich: W. Fink, 1974.

Schmidt di Friedberg, M. (ed.), *Élisée Reclus. Natura ed educazione* [Élisée Reclus. Nature and education]. Milan: Bruno Mondadori, 2007.

Scott, J.C., *The Art of Not Being Governed: An Anarchist History of Upland Southeast Asia*. New Haven: Yale University Press, 2009.

Scott Keltie, J., 'Obituary'. *The Geographical Journal* 57, no. 4(1921): 316–319.

Secord, J., 'Knowledge in Transit'. *Isis* 95, no. 4(2004): 654–672. doi:10.1086/430657.

Sempere, J., 'Decrecimiento y autocontención'. *Ecología Política* 35(2008): 35–44.

Sennett, R., *Flesh and Stone: The Body and the City in Western Civilization*. New York: Norton, 1996.

Seymour, J., *The Fat of the Land*. London: Faber & Faber, 1976.

Shackel, P.A. and B.J. Little, 'Post-processual Approaches to Meanings and Uses of Material Culture in Historical Archaeology'. *Historical Archaeology* 26, no. 3(1992): 5–11. doi:10.1007/BF03373538.

Shaffer, K., 'Tropical Libertarians: Anarchist Movements and Networks in the Caribbean, Southern United States and Mexico, 1890s–1920s'. In *Anarchism and Syndicalism in the Colonial and Postcolonial World, 1870–1940. The Praxis of National Liberation, Internationalism, and Social Revolution*, edited by S. Hirsch and L. van der Walt, 273–320. Leiden: Brill, 2013.

Shaffer, K., 'Latin Lines and Dots: Transnational Anarchism, Regional Networks, and Italian Libertarians in Latin America'. *Zapruder World* 1(2014) [Retrieved from http://www.zapruderworld.org/content/-r-shaffer-latin-lines-and-dots-transnational-anarchism-regional-networksand-italian].

Shaffer, K., 'Panama Red: Anarchist Politics and Transnational Networks in the Panama Canal Zone, 1904–1913'. In *In Defiance of Boundaries. Anarchism in Latin American History*, edited by G. de Laforcade and K. Shaffer, 48–71. Gainesville: University Press of Florida, 2015.

Shanks, M., 'Post-processual Archaeology and After'. In *Handbook of Archaeological Theories*, edited by R.A. Bentley, H.D.G. Maschner and C. Chippindale, 133–144. Lanham: AltaMira Press, 2008.

Shantz, J., *Against all Authority. Anarchism and the Literary Imagination*. Exeter: Imprint Academic, 2011.

Shantz, J., ed. *Specters of Anarchy. Literature and the Anarchist Literature*. New York: Algora Publishing, 2015.

Sharpe, S., Hynes, M. and R. Fagan, 'Beat Me, Whip Me, Spank Me, Just Make It Right Again: Beyond the Didactic Masochism of Global Resistance'. *Fibreculture* 6(2005): 16.

Shryock, R., 'Becoming Political: Symbolist Literature and the Third Republic'. *Nineteenth Century French Studies* 33, nos 3 & 4(1958): 385–398. doi:10.1353/ncf.2005.0037.

Shubin, A.V., *Nestor Machno: bandiera nera sull'Ucraina. Guerriglia libertaria e rivoluzione contadina (1917–1921)* [Nestor Makhno: black flag on Ukraine. Libertarian guerrillas and peasant revolution]. Milan: Elèuthera, 2012.

Silvey, R., 'Transnational Domestication: State Power and Indonesian Migrant Women in Saudi Arabia'. *Political Geography* 23, no. 3(2004): 245–264. doi:10.1016/j.polgeo.2003.12.015.

Sippel, A., 'Back to the Future: Today's and Tomorrow's Politics of Degrowth Economics (Décroissance) in Light of the Debate over Luxury among Eighteenth and Early Nineteenth Century Utopists'. *International Labor and Working-Class History* 75(2009): 13–29. doi:10.1017/S0147547909000039.

Sitrin, A. and D. Azzellini. *They Can't Represent Us: Reinventing Democracy from Greece to Occupy*. London and New York: Verso Books, 2014.

Sloterdijk, P., *Critique of Cynical Reason*. Minneapolis, MA: University of Minnesota Press, 1988.

Smith, A., 'Sardana, Zarzuela or Cake Walk? Nationalism and Internationalism in the Discourse, Practice and Culture of the Early Twentieth-Century Barcelona Labour Movement'. In *Nationalism and the Nation in the Iberian Peninsula. Competing and Conflicting Identities*, edited by C. Mar-Molinero and A. Smith, 171–190. Berg: Oxford, 1996.

Smith, A. (ed.), *Red Barcelona: Social Protest and Labour Mobilization in the Twentieth Century*. London: Routledge, 2003.

Smith, A., *Anarchism, Revolution and Reaction. Catalan Labour and the Crisis of the Spanish State, 1898–1923*. New York: Berghahn Books, 2007.

Sokolov, N.N., 'Pëtr Alekseevich Kropotkin kak geograf'. *Trudy Instituta Istoriia Estestvoznaniia AN SSSR* IV (1952): 408–442.

Sonn, R., *Anarchism and Cultural Politics in Fin de Siècle France*. Lincoln, NE: University of Nebraska Press, 1989.

Soper, K., 'Capitalocene'. *Radical Philosophy* 197(2016): 49–52.

Sorel, G., *Reflections on Violence*. Translated by J. Jennings. Cambridge: Cambridge University Press, 1999.

Şorman, A.H., 'Societal Metabolism'. In *Degrowth: A Vocabulary for a New Era*, edited by G. D'Alisa, F. Demaria, and G. Kallis, 41–44. New York: Routledge, 2015.

Soubeyran, O., *Imaginaire, science et discipline* [Imaginary, science and discipline]. Paris: Editions L'Harmattan, 1997.

Souza, M.L., 'Autogestão, autoplanejamento, autonomia: atualidade e dificuldades das práticas espaciais libertárias dos movimentos urbanos'. *Cidades: revista científica* 9, no. 15(2012): 59–94.

Souza, M.L., 'Panem et Circenses versus the Right to the City (Centre) in Rio de Janeiro: A Short Report'. *City* 16, no. 5(2012): 563–572. doi:10.1080/13604813.2012.709725.

Souza, M.L., 'Cidades Brasileiras, junho de 2013: o(s) sentido(s) da revolta' [Retrieved from http://passapalavra.info/2013/07/80798].

Souza, M.L., White, R.J. and S. Springer (eds), *Theories of Resistance: Anarchism, Geography and the Spirit of Revolt*. Lanham: Rowman & Littlefield, 2016.

Sparke, M., 'Political Geography: Political Geographies of Globalisation (2) – Governance'. *Progress in Human Geography* 30, no. 3(2006): 357–372. doi:10.1177/0309132507086878.

Springer, S., 'Anarchism and Geography. A Brief Genealogy of Anarchist Geographies'. *Geography Compass* 7, no. 1(2013): 46–60. doi:10.1111/gec3.12022.

Springer, S., 'War and Pieces'. *Space and Polity* 18, no. 1(2014): 85–96. doi:10.1080/13562576.2013.878430.

Springer, S., 'Why a Radical Geography Must Be Anarchist'. *Dialogues in Human Geography* 4, no. 3(2014): 249–270. doi:10.1177/2043820614540851.

Springer, S., 'The Limits to Marx. David Harvey and the Condition of Postfraternity'. (2015) [Retrieved from https://www.academia.edu/12638612/The_limits_to_Marx_David_Harvey_and_the_ condition_of_postfraternity on 10 July 2016].

Springer, S., *The Anarchist Roots of Geography: Toward Spatial Emancipation*. Minneapolis: Minnesota University Press, 2016.

Springer, S., Ince, A., Pickerill, J., Brown, G. and A. Barker, 'Reanimating Anarchist Geographies: A New Burst of Colour'. *Antipode*44, no. 5(2012): 1591–1604. doi:10.1111/j.1467–8330.2012.01038.x.

Springer, S., White, R.J. and M.L. Souza (eds), *The Radicalization of Pedagogy: Anarchism, Geography and the Spirit of Revolt*. Lanham: Rowman & Littlefield, 2016.

Squatting Europe Collective, *Squatting in Europe: Radical Spaces, Urban Struggles*. New York: Autonomedia, 2013.

Staeheli, L.A., Ehrkamp, P., Leitner, H. and C.R. Nagel, 'Dreaming the Ordinary: Daily Life and the Complex Geographies of Citizenship'. *Progress in Human Geography* 36, no. 5(2012): 628–644. doi:10.1177/0309132511435001.

Stafford, D., *From Anarchism to Reformism: A Study of the Political Activities of Paul Brousse with the First International and the French Socialist Movement, 1870–1890*. London: Weidenfeld & Nicolson, 1971.

Stansell, C., *American Moderns. Bohemian New York and the Creation of a New Century*. New York: Metropolitan Books, 2000.

Stears, M., 'Guild Socialism'. In *Modern Pluralism. Anglo-American Debates since 1880*, edited by M. Bevir, 40–59. Cambridge: Cambridge University Press.

Steffen, W., Crutzen, P.J. and J.R. McNeill, 'The Anthropocene: Are Humans now Overwhelming the Great Forces of Nature?' *Ambio* 36, no. 8(2007): 614–621. doi:10.1579/0044-7447(2007)36[614:TAAHNO]2.0.CO;2.

Stoddart, D.R., 'The RGS and the Foundations of Geography at Cambridge'. *The Geographical Journal* 141, no. 2(1975): 216–239.

Strathern, M., *Partial Connections*. Walnut Creek: AltaMira, 2004.

Struthers, D., '"The Boss has no Color Line": Race, Solidarity, and a Culture of Affinity in Los Angeles and the Borderlands, 1907–1915'. *Journal of the Study of Radicalism* 7, no. 2(2013): 61–92. doi:10.14321/jstudradi.7.2.0061.

Subcomandante Marcos, 'Los sueños buenos y malos'. (2006) [Recording retrieved from https://soundcloud.com/gblas/los-cojolites-los-sue-os?in=jorge-herrera-carrasco/sets/rebeldia].

Tilly, C., 'War Making and State Making as Organised Crime'. In *Bringing the State Back In*, edited by P. Evans, D. Rueschemeyer and T. Skocpol, 169–187. Cambridge: Cambridge University Press, 1985.

Todes, D., *Darwin without Malthus. The Struggle for Existence in Russian Evolutionary Thought*. New York and Oxford: Oxford University Press, 1989.

Toledo, V.M. and N. Barrera-Bassols, *La memoria biocultural. La importancia ecológica de las sabidurías tradicionales* [Biocultural memory. The ecological importance of traditional wisdoms]. Popayán, Colombia: Universidad del Cauca, 2014.

Tomchuk, T., *Transnational Radicals. Italian Anarchists in Canada and the U.S. 1915–1940*. Winnipeg: University of Manitoba Press, 2015.

Toro, F., 'Una visión crítica de la sostenibilidad y algunas reflexiones para salir del imaginario dominante' [A critical vision of sustainability and some reflections on leaving the dominant imaginary]. In *¿Otro municipio es posible? Guanabacoa en La Habana* [Is another municipality possible? Guanabacoa in La Habana], edited by Y. Farrés and A. Matarán (eds), 32–63. Granada: Atrapasueños, 2012.

Toro, F., 'La sensibilidad ecológica en el pensamiento de Elisée Reclus: Una revisión necesaria para entender las claves de la crisis ecológica contemporánea' [Ecological sensitivity in the thinking of Elisée Reclus: A necessary review to understand the keys to the contemporary ecological crisis]. In *Relación entre la sociedad y el medio ambiente en la geografía moderna* [The relationship between society and the environment in modern geography], edited by M. Frolova, 181–204. Granada: Editorial Universidad de Granada, 2015.

Toro, F., 'Educating for Earth Consciousness. Ecopedagogy within Early Anarchist Geography'. In *The Radicalization of Pedagogy. Anarchism, Geography, and the Spirit of Revolt*, edited by S. Springer, M.L. de Souza and R.J. White, 193–221. London and New York: Rowman & Littlefield, 2016.

Torúa Cienfuegos, A., *El magonismo en Sonora (1906–1908). Historia de una persecución* [The magonismo in Sonora (1906–1908). History of a persecution]. México: Ed. Hormiga Libertaria y Nosotros, 2010.

Trainer, T., 'The Degrowth Movement from the Perspective of the Simpler Way'. *Capitalism, Nature, Socialism* 26, no. 2(2015): 58–75. doi:10.1080/10455752.2014.987150.

Turcato, D., 'Italian Anarchism as a Transnational Movement, 1885–1915'. *International Review of Social History* 52, no. 3(2007): 407–444. doi:10.1017/s0020859007003057.

Turcato, D., 'Nations without Borders: Anarchists and National Identity'. In *Reassessing the Transnational Turn: Scales of Analysis in Anarchist and Syndicalist Studies*, edited by C. Bantman and B. Altena, 25–42. New York: Routledge, 2015.

Turcato, D., *Making Sense of Anarchism: Errico Malatesta's Experiments with Revolution, 1889–1900*. Oakland: AK Press, 2015.

Van der Steen, B., Katzeff, A. and L. Van Hoogenhuijze, *The City is Ours: Squatting and Autonomous Movements in Europe from 1970s to the Present*. Oakland: PM Press, 2014.

Vincent, K.S., *Between Marxism and Anarchism: Benoît Malon and French Reformist Socialism*. Berkeley: University of California Press, 1992.

Virno, P., *Gramática da multidão. Para uma análise das formas de vida contemporâneas* [Grammar of crowds. For an analysis of contemporary ways of life]. São Paulo: Annablume, 2013.

Vucinich, A., *Darwin in Russian Thought*. Berkeley: University of California Press, 1988.

Vuilleumier, M., *Histoire et combats: Mouvement ouvrier et socialisme en Suisse, 1864–1960* [History and battles: labour movements and socialism in Switzerland]. Geneva: Collège du Travail, 2012.

Ward, C., *Anarchy in Action*. London: Freedom Press, 1973.

Ward, C., *Housing: An Anarchist Approach*. London: Freedom Press, 1976.

Ward, C., *Child in the City*. London: Architectural Press, 1977.

Ward, C., *Chartres: The Making of a Miracle*. London: Folio Society, 1986.

Ward, C., *The Child in the Country*. London: NCVO Publications, 1988.

Ward, C., *Influences: Voices of Creative Dissent*. Hartland, Devon: Green Books, Hartland, 1991.

Ward, C., *Anarchism: A Very Short Introduction*. Oxford: Oxford University Press, 2004.

Ward, C., *Talking Green*. Nottingham: Five Leaves Books, 2012.

Ward, C. and D. Hardy, *Goodnight Campers! The History of the British Holiday Camp*. London: Mansell, 1986.

Ward, D., 'Alchemy in Clarens. Kropotkin and Reclus, 1877–1881'. In *New Perspectives on Anarchism*, edited by N. Jun and S. Wahl, 209–226. Lanham: Lexington Books, 2010.

Watkins, P., *La commune (Paris, 1871)*. France, 2000. 345 min (DVD).

Watts, A., *Become What You Are*. Boston, MA: Shambhala Publications, 2003.

Weaver, S., 'Liquid Racism and the Danish Prophet Muhammad Cartoons'. *Current Sociology* 58, no. 5(2010): 675–692. doi:10.1177/0011392110372728.

Weaver, S., *The Rhetoric of Racist Humour: US, UK and Global Race Joking*. Farnham: Ashgate, 2011.

Weisberg, G. (ed.), *Montmartre and the Making of Mass Culture*. New Brunswick, NJ: University of Rutgers Press, 2001.

White, R.J., Springer, S. and M.L. Souza (eds), *The Practice of Freedom: Anarchism, Geography and the Spirit of Revolt*. Lanham: Rowman & Littlefield, 2016.

White, D.F. and G. Kossoff, 'Anarchism, Libertarianism and Environmentalism: Anti-Authoritarian Thought and the Search for Self-Organizing Societies'. In *The SAGE*

Handbook of Environment and Society, edited by J. Pretty, A.S. Ball, T. Benton, J.S. Guivant, D.R. Lee, D. Orr, M.J. Pfeffer and H. Ward, 50–65. London: SAGE Publications 2008.

Wilbert, C., and D. White, *Autonomy, Solidarity, Possibility. The Colin Ward Reader.* Edinburgh: AK Press, 2011.

Williams, E., 'Signs of Anarchy: Aesthetics, Politics, and the Symbolist Critic at the Mercure de France, 1890–1895'. *French Forum* 29, no. 1(2004): 45–68. doi:10.1353/frf.2004.0039.

Wilson, C., *Paris and the Commune, 1871–78: The Politics of Forgetting.* Manchester: Manchester University Press, 2007.

Wilson, E., *Bohemians: The Glorious Outcasts.* London: I.B. Tauris, 2000.

Woehrlin, W., *Chernyshevskii. The Man and the Journalist.* Cambridge, MA: Harvard University Press, 1971.

Wolff, L., *Inventing Eastern Europe: The Map of Civilization in the Mind of the Enlightenment.* Stanford: Stanford University Press, 1994.

Woodcock, G. and I. Avakumović, *The Anarchist Prince: A Biographical Study of Peter Kropotkin.* London: Boardman Books, 1950.

Woodcock, G., *Anarchism. A History of Libertarian Ideas and Movements.* Cleveland, Ohio: The World Publishing Company, 1962.

Woodward, K. and M. Bruzzone, 'Touching like a State'. *Antipode* 47, no. 2(2015): 539–556. doi:10.1111/anti.12119.

Wurst, L.A., 'The Historical Archaeology of Capitalist Dispossession'. *Capital and Class* 39, no. 1(2015): 33–49. doi:10.1177/0309816814564131.

Yoffee, N., *Myths of the Archaic State: Evolution of the Earliest Cities, States, and Civilisations.* Cambridge: Cambridge University Press, 2005.

Yoffee, N., 'The Power of Infrastructures: A Counternarrative and a Speculation'. *Journal of Archaeological Methodology and Theory.* doi:10.1007/s10816–10015–9260–0 [forthcoming].

Zimmer, K., *'The Whole World Is Our Country': Immigration and Anarchism in the United States, 1885–1940*, doctoral thesis, University of Pittsburgh, 2010.

Zimmer, K., 'A Golden Gate of Anarchy: Local and Transnational Dimensions of Anarchism in San Francisco, 1880s-1930s'. In *Reassessing the Transnational Turn. Scales of Analysis in Anarchist and Syndicalist Studies*, edited by C. Bantman and B. Altena, 100–117. London: Routledge, 2015.

Zimmer, K., *Immigrants against the State. Yiddish and Italian Anarchism in America.* Urbana: University of Illinois Press, 2015.

Index

References to illustrations are in **bold**

Adam, Paul 77–8
Adamovsky, Ezequiel 117
Adorno, Theodor 67, 188
Agnew, J. 180
Aldred, Guy 115
allotments, Ward on 160–1
Alonso González, P. 189, 190; on archaeology 187
anarchism xi; anarchy, distinction xii–xiii; Barcelona (1936) 2, 11, 19–21; biographical approaches 137; and the city 7–9, art in 16–19; and decentralisation 197–8; definition 138; diversity 198, 203–5; and ecology 92; and the Futurists 18; geography, relationship 131, 136–8, 142, 145; and the Global South 13–16; and humour 65–8, 73–4, 76; importance of Switzerland 113; indigenous 204–5; landscapes 196–9; Montmartre 65; and nation 114, 120–1, 125; and nihilism 77, 142; nostalgia for primitivism 79–80, **80**, 81; 'Other' geographies perspective 195, 199–203, 206; as possibilities 195–6; principles 197–8; purpose xii; and republicanism 114–15; roots 114; as science 141, 142; spatial dimension 143; Ward 159–60, 161–2; Western perspective 195, 197–9, 203
anarchisms 7
anarchist geographers 113; networks 130; variant visions 139
anarchist geographies xi, 117, 199; as anti-geopolitics 136; concept 134; genealogy 142–4; problems with term 132; reception 134; rediscovery of 1, 146

anarchist periodicals, Italy and USA 46–7
anarchist periodicals (USA): cash flow 52; circulation 51; distribution by region 55; distribution by state 41; donations 52; distribution by state 56; per capita by state 58–9; editors 59–61; expenditure 52; income 52; lifespan by city 43; lifespan by state 42–3; sales 52; subject matter 60–1; subscriptions 52; transnationalism 61
anarchist press: *see* anarchist periodicals (USA); Italian anarchist press (USA)
anarchists: anti-colonialism 115; and Messina Earthquake (1908) 34–5, 38; migration to USA 42, 44; Spanish Civil War 14; support to national liberation movements 115; in *The Secret Agent* (Conrad) 77
anarchy: anarchism, distinction xii–xiii; nature of xi
Anarchy journal 153, 158
Anderson, Benedict 15, 26, 115
Anderson, Margaret 17
Anthropocene 195, 202, 203; definition 207n4; multiple versions 205
anti-colonialism: anarchists 115; Reclus 105–6, 139–40
anti-utilitarianism: of Reclus 96
Antipode journal 129, 139
Aranha, José Pereira de Graça 124
archaeology: Alonso on 187; definition 181; as metaphor 179; origins of discipline 187; post-processual 182; scope 181–2; and the state 183, 184, 185–6, 189, 191; task of 188
archy xii

Milton Keynes UK
Ingram Content Group UK Ltd.
UKHW040105071024
449327UK00019B/831